Biosphere Reserves in the Mountains of the World

Excellence in the Clouds?

Austrian MAB Committee (ed.)

ÖAW
Austrian Academy
of Sciences

United Nations
Educational, Scientific and
Cultural Organization

Man and
the Biosphere
Programme

Content

ALPEC	Alianza para Ecosistemas Criticos
AMA	Andean Mountains Association
BCBR	Bilateral Council of Krkonose/Karkonosze BR
BR	Biosphere Reserve
BRIM	Biosphere Reserve Integrated Monitoring
CBD	Convention on Biological Diversity
CBR	Carpathian Biosphere Reserve
CBRA	Canadian Biosphere Reserves Association
CEA	Critical Ecosystem Alliance
CONANP	National Commission of Natural Protected Areas, Mexico
COP	Conference of the Parties
CPAMETT	Carpathian Protected Area Management Effectiveness Tracking Tool
DSW	German Foundation for World Population (Deutsche Stiftung Weltbevölkerung)
EC	European Commission
ECNC	European Centre for Nature Conservation
EEA	European Environment Agency
eea®	European Energy Award®
EFAP	Ethiopian Forestry Action Programme
ETC/BD	European Topic Centre on Biological Diversity
EU	European Union
FAO	Food and Agriculture Organisation of the United Nations
FDI	Foreign Direct Investment
FWF	Austrian fund for scientific research projects
GEF	Global Environmental Facility
GESGIAP	Sierra Gorda Ecological Group
GIS	Geographical Information System
GLOCHAMORE	Global Change in Mountain Regions
GLOCHAMOST	Global Climate Change in Mountain Sites, Coping Strategies for Mountain Biosphere Reserves
GLORIA	Global Observation Research Initiative in Alpine Environments
GPS	Global Positioning System
GTZ	German Association for Technical Cooperation in Developing Countries (since 2011: GIZ)
IBP	International Biological Programme
ICC	International Coordination Council of the Man and the Biosphere Programme
IBA	Important Bird Areas
IHP	International Hydrological Programme
IPCC	Intergovernmental Panel on Climate Change
IUCN	International Union for Conservation of Nature
LTER	Long Term Ecological Research (LTER)
MAB	UNESCO's Man and the Biosphere Programme
MAB-NC	MAB National Committee
MAP	Madrid Action Plan (adopted 2008 during the Third World Congress of Biosphere Reserves)
METT	Management Effectiveness Tracking Tool (developed by WWF and the World Bank)
MRI	Mountain Research Initiative
NABU	Naturschutzbund Deutschland
Nearctic	The Nearctic is one of the eight terrestrial ecozones dividing the Earth's land surface. It covers most of North America, including Greenland and the highlands of Mexico.
NGO	Non-Governmental Organisation
NP	National Park
NTFP	Non-Timber Forest Products
PA	Protected Areas
PCDA	Programme for Conservation and Development of the Arganeraie
PFM	Participatory Forest Management
PGI	Protected Geographical Indication
PoWPA	CBD's Programme of Work on Protected Areas
PPP	Public-Private Partnership
RARBA	Arganeraie Biosphere Reserve Association's Network
TANAP	Tatry National Park
UNCCD	United Nations Convention to Combat Desertification
UNCED	United Nations Conference on Environment & Development, hold in 1992 in Rio de Janeiro (also referred to as 'Earth Summit')
UNEP	United Nations Environment Programme
UNDP	United Nations Development Programme
UNESCO	United Nations Educational, Scientific and Cultural Organisation
WCMC	UNEP's World Conservation Monitoring Centre
WDPA	World Database on Protected Areas
WNBR	World Network of Biosphere Reserves
WWF	World Wide Fund for Nature

by Thomas Schaaf, UNESCO

Mountain landscapes are among the most complex and fragile ecosystems on earth. Their very verticality and exposure to the sun and prevailing wind directions produce a range of different habitats whose composition may vary dramatically with minor variations in altitude. These differences can be obvious in the tropics or sub-tropics – such as the presence of palm trees at lower altitudes and glaciers at higher ones – or more subtle, such as the shifts in insect species as one moves up a mountain slope. Anthropogenically induced erosion (e.g. over-grazing) and natural hazards (floods, avalanches, glacial lake outbursts, seismic activities) testify to the fragility of mountain environments.

And yet the fragility and in particular the diversity of mountains make them most valuable from the point of view of biodiversity conservation. Relict species from former climatic periods and rare and endangered species find refuge in relatively inaccessible areas that benefit from low levels of disturbance by humans. However, biodiversity conservation is at risk when economic development, social development and environmental protection are practised in a disaggregated manner. It should not be overlooked that mountains also provide homes to people and their distinct cultures. The rational use of mountain resources (water, timber, minerals, medicinal plants etc.) is mandatory in a world marked by rapid demographic growth and accelerating global change including climate change.

It is not an easy task to combine the needs for biodiversity conservation in mountains with promoting future-oriented life-styles for people, thus fostering sustainable development for mountain communities. But this is precisely the objective of the biosphere reserve concept which was developed by UNESCO's Man and the Biosphere (MAB) Programme. Biosphere reserves are areas of terrestrial ecosystems promoting solutions to reconcile the conservation of biodiversity with sustainable use by means of an integrated land-use system. They are internationally recognised, nominated by national governments and remain under the sovereign jurisdiction of the states where they are located. Biosphere reserves serve in some ways as living laboratories for testing and demonstrating the integrated management of land, water and biodiversity. Collectively, biosphere reserves form a world network: the World Network of Biosphere Reserves (WNBR). Within this network, exchanges of information, experience and staff are facilitated. Currently (2010), there are over 564 biosphere reserves in 109 countries, and many of them are located in mountains.

In this publication you will find various examples of mountain biosphere reserves from all over the world – and read about the important roles they play as sites for conserving biodiversity, international scientific collaboration and sustainable use of natural resources in line with conservation objectives.

Regarding the sustainable use of natural resources, organic farming, ecotourism and the labelling of regional quality products hold great promise for mountain communities and environments alike. One of the major themes to be addressed by the Rio+20 conference to be held in Brazil in 2012 (twenty years after the Earth Summit) will be 'green economy in the context of sustainable development and poverty reduction'. As stated in this publication, some mountain biosphere reserves illustrate this theme very well.

In the field of scientific collaboration, the diversity and fragility of mountain ecosystems make mountains excellent indicator sites to assess, study and monitor global climate change impacts on mountain environments as well as on the socio-economic livelihoods of mountain people. This was the starting point for the project 'Global Change in Mountain Regions' (GLOCHAMORE) funded by the European Commission, sponsored by UNESCO and run under the leadership of the University of Vienna (Austria) in an attempt to detect signals of global climate change in mountain ranges the world over. Since 2003, some 300 scientists have participated in the GLOCHAMORE project which encompasses a wide range of different scientific disciplines. The project also served to create a network for studying and monitoring sites in over 20 mountain biosphere reserves which represent the world's major mountain ranges. As a result of the project, the GLOCHAMORE Research Strategy was established. It recommends specific actions to be taken in order to detect and monitor signals of global climate change in mountain biosphere reserves. In the implementation of the Research Strategy, UNESCO's follow-up project GLOCHAMOST notably addresses the main axes of causality for global change in mountain biosphere reserves: climate, land-use change, biodiversity, and mountain economies (cf. the example of Katunskiy BR in this publication). In view of their rich biodiversity, it is no wonder that many mountain biosphere reserves are known as biological hotspots. The conservation of species and their habitats is at the core of each biosphere reserve where strictly protected areas ensure the long-term survival of flora and fauna, including rare and endangered species.

Preface

UNESCO's Man and the Biosphere (MAB) Programme is very grateful to the Austrian MAB National Committee under the auspices of the Austrian Academy of Sciences for having commissioned the preparation of this publication. This publication is a most welcome contribution of the Austrian MAB National Committee to commemorate the 40th Anniversary of the MAB Programme, which was officially launched in 1971. Our particular thanks go to Georg Grabherr, Chair of the Austrian MAB National Committee, Günter Köck, Secretary-General of the Austrian MAB National Committee, and Sigrun Lange, Executive Secretary of E.C.O. Deutschland, as well as to all authors who contributed to this publication. We hope that this book will arouse your interest. It shows that biosphere reserves can only function properly if people are closely involved with the appropriate management of their environment. Wherever they achieve this, mountain biosphere reserves will be noted for 'excellence in the clouds'.

Dr. Thomas Schaaf
Chief, Section for Ecological Sciences and Biodiversity
Division of Ecological and Earth Sciences
UNESCO Man and the Biosphere (MAB) Programme

Mountain Regions of the World

Threats and Potentials for Conservation
and Sustainable Use

Some expert's opinions

Mountains are found on every continent: Tuxer Alps in Austria (© Sigrun Lange).

An Overview of the World's Mountain Environments

by Georg Grabherr & Bruno Messerli

Mountains are found on every continent, from the equator polewards as far as land extends. In their entirety – because of their three-dimensional nature, as a single great landscape category or ecosystem in the broadest sense – they encompass the most extensive array of topography, geology, climate, flora and fauna, as well as human cultural differentiation that is known to humankind. They cover, depending on which definition is applied, between 22 and 25 per cent of the Earth's total land area.

To a naturalist, mountains are primarily hotspots of biodiversity. This is due to their vertical expansion which creates different climatic conditions only short distances apart. In addition, a heterogeneous relief gives rise to a mosaic of different habitats close to each other, which adds to the great landscape diversity of mountain terrain. The latter is particularly evident at and beyond tree lines where zonal grasslands or dwarf-shrub heaths can alternate with snow beds, rock faces and wetlands. Furthermore, mountains are steep. Morphodynamic processes related to high-relief energy create specific habitats such as screes, landslides and avalanches.

They host a flora and fauna adapted to recurrent disturbance. Overall, mountain regions that range from tropical and sub-tropical to temperate climate, host – on a square of 100 x 100 kilometres – more than 5,000 vascular plant species (Himalayas, tropical Andes), while temperate mountains can host up to 3,000 species (Alps), and boreal mountains have at least twice as many species as the surrounding lowlands. Combined with habitat diversity, this high species diversity relates to an exceptional richness in biotic communities. So far, more than 600 floristically defined types (associations) have been described for the Alps. Biosphere reserves should be designated in order to protect this wealth of diversity at the same time as promoting sustainable use.

From an anthropocentric point of view, mountains are home to a significant part of the world's population. This figure was estimated by FAO for the UN International Year of Mountains 2002 to be approx. twelve per cent. If you include the rural and especially the urban population that lives in contact with mountain environments in the immediately adjacent lowlands, then this amount increases to more than 25 per cent of the global population. For about 50 per cent who live outside mountain areas, they serve as water towers, in addition to offering a variety of natural products, and they are important for recreation. Mountain dwellers have developed an array of very different cultures even in one particular mountain system. The intensity of human interventions, however, depends on the relationship between the position of a certain mountain system and its specific life zones. Cultivated fields can be found at altitudes of up to 4,000 metres in the tropical Andes and in the Alps up to 1,800 metres, whereas they are absent from boreal mountains. Generally, mountains in their upper regions are characterised by harsh environments. Therefore, many of these montane and higher zones have remained in a natural, sometimes even pristine state. Mountain biosphere reserves reflect this pattern. Some are wildernesses where maintaining this status is the main goal, others focus on the cultural aspects of reconciling development with traditional land-use practices and values.

In 1992, during and after the Earth Summit in Rio de Janeiro, the world's mountains were given attention at the highest political level. This had its formal expression in the inclusion in Agenda 21, of chapter 13 on 'Managing Fragile Ecosystems: Sustainable Mountain Development'. No doubt, this produced heightened awareness of the essential role played by mountains as an integral part of the global biophysical and socioeconomic system. Now we are close to the next global conference, 20 years after the 1992 Rio Conference. It will take place in 2012, again in Rio de Janeiro, where the scientific mountain community has to demonstrate its progress and present the results of the past two decades. In this connection it is probably of interest to read the introductory article to the last resolution adopted by the UN General Assembly on 11th March 2010 on the subject of 'Sustainable mountain development': '… notes with appreciation that a growing network of governments, organisations, major groups and individuals around the world recognises the global importance of mountains as the source of most of the Earth's freshwater, as repositories of rich biological diversity and other natural resources, including timber and minerals, as providers of some sources of renewable energy, as popular destinations for recreation and tourism and as areas of important cultural diversity, knowledge and heritage, all of which generate positive, unaccounted economic benefits'.

Mountains of the world

In terms of climate, geology, vegetation, animal life, and human interference, each mountain environment is somewhat unique. Nevertheless, both geographers and ecologists have attempted to classify mountains by identifying similarities or dissimilarities. The various sets of criteria used depend on the focus of interest:

• geological (period of orogenesis, lithological),
• ecological and climatological (position in and in relation to life zones; seasonality/precipitation/air density, solar radiation),
• biogeographical (position in and in relation to floristic/ faunistic realms),
• cultural, social and economic (diversity of landscapes and land-use systems).

Here, we combine various life-zone systems with mountain regions in order to give an overview from an ecological point of view, also adding some information on biogeographical aspects as well as human interventions and impacts.

Arctic mountains

Arctic mountains are located north of the arctic treeline such as the Byrranga Mountains on the Taymir peninsula – the only place on the globe with a continuous terrestrial gradient from the taiga region to the polar desert. Mountains on arctic islands are, for example, those on Svalbard, Novaya Zemlya, Franz Josef Land and on the islands of the Canadian archipelago. Greenland, the largest biosphere reserve on earth, is outstanding in terms of size and latitudinal expansion. Even at the base of arctic mountains mean annual temperatures are below zero degree Celsius, the growing season is short and annual precipitation often less than 300 millimetres. Very low evaporation, however, results in a positive water balance. Rising from the arctic tundra or even the polar desert (e.g. Franz Josef Land) the elevational gradient is steep, and it is not possible to distinguish between different zones. Dwarf shrubs, fell-fields dominated by lichens and mosses and/or wet tundra are typical formations. Most of these mountains are covered in vast ice sheets. In ice-free habitats, cryoturbation keeps places free from rooting plants. In the arctic, vascular plants belong to the holarctic realm throughout, and therefore their diversity is low. On the other hand, the arctic is rich in animal life. Migrating birds are numerous, benefiting from resources in the sea or the enormous production of insects provided mainly by the wet tundra. Polar bears use remote places such as Franz Josef Land for reproduction.

Mountains of the arctic are among the most hostile environments for humans, and therefore remained in an almost pristine state. Even the first nations such as the Inuit were not known for making use of the upper reaches of arctic mountains. They are, however, places of increasing scientific interest, particularly for the study of climate change impacts. Antarctic mountains – those on the mainland and those on the islands (e.g. South Georgia) – might be considered comparable to arctic mountains. They differ insofar, however, as they belong to the Antarctic realm with its tussock grasses. Only two vascular plants occur in continental Antarctica. By contrast, as in the Arctic, animal life is very specific and rich, with isolated groups such as penguins.

Boreal mountain regions

Boreal mountains cover huge areas, i.e. approx. five per cent of the Earth's land surface. They rise from the endless coniferous forests (taiga) north of 50 degrees latitude. Exceptions are oceanic mountains (Northern Scandes, mountain of Beringia) where deciduous forests or shrublands dominate (e.g. *Betula*, *Alnus*). In many of these mountains, two elevational zones can be distinguished, the forest zone, and beyond the tree line, an alpine zone. The tree line often extends over several hundreds of metres transgrading into dwarf scrub heath mixtures of willows and Ericaceae. Higher up sedges and rushes form more or less closed swards, while fell-fields are characteristically found on wind-swept ridges. Typical fell-field plants such as the cushion plants *Silene excapa* or *Saxifraga oppositifolia* are absolutely frost-resistant and therefore able to withstand extreme frosts (experimentally even liquid air) typical of boreal alpine environments. Frost action causes vast boulder fields (goltsy), predominantly in continental mountains. Cryoturbation, in particular solifluction, produces patterned ground which reduces the availability of safe sites for rooting plants. High mountains such as those of the Canadian Rockies reach up into the nival zone where only a few plants and animals can survive. Not quite so high are the Caledonian mountains of northern Europe, the northern and polar Ural, Sibirian ranges such as the Putorana, the Verkhoyansk, the Kamchatka (many volcanoes) and Anadyr mountains as well as the northern Altai. The climate of these mountain ranges is strongly seasonal with long winters, a short vegetation period (the monthly mean in the warmest month is approx. 10°C), low precipitation, most of it as snow. A similar climate is found in the southernmost Andes of Patagonia. The forests, however, are composed of broad-leaved trees (mainly *Nothofagus antarctica*).

Polar desert at Ziegler Island, Franz Josel Land, Russia (© Harald Pauli).

Boreal mountains are hostile environments and therefore only scarcely populated. Some indigenous first nations, most of them nomadic reindeer farmers (restricted to Eurasia) used the transitional areas from montane to alpine, according to the seasons. Further activities have been fur hunting and fishing along the coastlines. Oil industry, mining, and logging have produced some disastrous impacts on the vast wilderness. The forests, some of them approaching the alpine zone, are characterised by more of less undisturbed wildlife.

Temperate mountains

These are the mountains of middle latitudes where the zonal climate is strongly seasonal with frosty winters causing dormancy. Summers, however, are reasonably warm; monthly mean values for July and August may rise up to approx. +20 degrees Celsius. Annual averages range from 6 up to 10 degrees Celsius. Where precipitation is more than 400 millimetres annually (mainly summer rain; and snow in winter), the related zonal vegetation is a deciduous forest. The shedding of leaves in autumn avoids frost drought in late winter. Where the precipitation values are lower, steppes and prairies, or even deserts occupy vast areas in continental Eurasia and Northern America. In both subregions, some of the most spectacular mountain environments reach up far into the glaciated nival zone (Cascades, Rocky Mountains, Alps, Caucasus, Tienshan, Southern Altai). Above 6,000 metres, they reach into the so-called aeolian zone where only soil microbes and algae survive, and wind-transported arthropods develop simple food chains. Outstanding in this respect are the Himalayas, but also the highest peaks in the Tienshan. Southern-hemisphere mountain environments which could be considered temperate, are the New Zealand Alps and the Andes of Chile and Argentina ranging from 35 to 45 degrees south.

All these mountains show a pronounced elevational zonation with deciduous forests in the lowlands or with steppes, semi-deserts, or deserts in the most continental regions. As precipitation increases with altitude, forests can grow in the dry mountains at middle elevations, but they are almost all coniferous forests. Coniferous forests are also characteristic for the so-called montane zone of oceanic or suboceanic mountains. In all these mountains a low-temperature related upper tree line or tree line ecotone demarcates the transition into the alpine zone – by definition treeless. Dwarf-shrub heath, and higher up sedge heath (*Carex, Kobresia*) represent the zonal vegetation, whereas in moist and wet regions snow beds and fens are interspersed. There is high floristic similarity among the northern mountains. The North American Rocky Mountains belong to the holarctic realm and the same is true for the Alps or the Himalayas. This is different in the south. The tussock grasslands of the alpine zone of the New Zealand Alps, for example, host taxa typical of the Australian realm, whereas those of the temperate Andes are neotropical or antarctic. In both regions, however, species of *Nothofagus*, an old Gondwana relict, dominate the montane forests; remarkably, that of New Zealand is exclusively evergreen.

Some of the temperate mountain regions have been settled since prehistoric times (Alps, Pyrenees, Himalayas). The best evidence for this is the ice mummy (Ötzi) found in the Tyrolean Alps. Others are still impressive wildernesses (Japanese Alps, Rocky Mountains, Cascades, Patagonian Andes, Alps of New Zealand). In the Eurasian mountains typical forms of trans-humance systems have been practised with summer pastures on high ground and winter pasturing in lowland steppes. Oceanic or suboceanic temperate mountains receive high amounts of snow in winter. Permanent settlement demands storage of food for people and fodder (hay) for livestock. The response of the mountain people was to develop meadow cultures combined with summer pasturing on high ground (Alps, Carpathians, Pyrenees, Caucasus). Some of these mountains which have been settled for thousands of years, have undergone a dramatic transformation with traditional agriculture declining and new activities such as tourism (Alps, Pyrenees) expanding. Others are still intact subsistent farming regions (Himalayas), those formerly collectivised are in a state of somewhat chaotic development (Tienshan, Altai). According to the great variety of temperate mountain regions in terms of ecological character as well as human intervention, biosphere reserves can serve different interests, namely those of conservation and of economic development or something in between.

Representative of the old Alp-culture of the European temperate mountains: Herdsman in Kleinwalsertal, Austria, providing salt for cattle (© Hans Grabherr).

Subtropical mountains

This type of mountains includes those belonging to the Mediterranean and hot desert life zones. Mediterranean climate is strongly seasonal both in terms of temperature and precipitation. Winters are fairly cold and rainy with much snow in the mountains. Frosts still act as selective environmental filter for plants and animals. Summers are hot and dry all the way up to high ground. The annual mean temperature is above +10 degrees Celsius.

Mountains with an alpine zone (or oromediterranean) are mainly those surrounding the Mediterranean Sea, the Californian Sierra Nevada, the Snowy Mountains in Australia, and the Andes of central Chile between 33 and 35 degrees latitude. A few approach the nival zone (Zagros, Elburz, Hindukush). Evergreen sclerophyllous forests and shrublands are the zonal vegetation types, higher up replaced by deciduous forests. Beyond the tree line ecotone, thorny cushion communities hosting numerous geophytes (*Tulipa*, *Iris*, *Scilla*) are peculiar to Mediterranean mountains. There is, however, no other life zone where the consideration of biogeographical aspects is more important. In Australia for example, Eucalyptus dominates all habitats where trees can grow. Even the tree line in the Snowy Mountains is formed by the so-called snow gum (*Eucalyptus niphophila*). Columnar cacti are intermixed in the Andean montane sclero-phyllous forest with species linked to the neotropical realm. Although the Californian and Mediterranean mountains have many genera in common, the peculiar thorny cushions are restricted to the latter.

Between 22 and 25 degrees south, the Andes, and between 20 to 24 degrees north the mountains of the central Sahara (Tibesti and Hoggar) rise up from true hot deserts. Even at higher elevations, precipitation is too low for forests to grow. Steppes or xeromorphic shrublands are the zonal vegetation type. Higher up it becomes drier again and some kind of cold semideserts have developed, while in the Andes the exceptionally dense cushions of the genus *Azorella* thrive.

The Mediterranean life zone is one of the best environments for humans. The earliest advanced urban cultures developed here. Most natural forests in the lowlands, middle and lower slopes have been replaced by cultivated land. The high ground has been used for pasturing. This is true for the mountains surrounding the Mediterranean Sea, but completely different in other regions. Until the appearance of Europeans, the first nations had developed sustainable forms of using the natural resources. The Californian Indians, for instance, harvested acorns and hunted by means of fire. Aborigenes came to the alpine areas of the Snowy Mountains using the mass migration of the bogong moth (*Agrotis infusa*). The Europeans introduced land-use practices such as pasturing for which the native plants were not suitable. Many species from Europe had been introduced resulting in a high proportion of neophytes in the recent flora. This is less evident in the mountains. Nowadays, the abandonment of mountain farmland is seen as a major

Top: Subtropical-Mediterranean spiny cushion vegetation in Sierra Nevada BR, Spain (© Georg Grabherr).

Bottom: The giant rosette formations of Lobelia rhynchocephala are typical for the seasonal climate in the Alpine-nival habitats of the tropical and subtropical life zone. The photo was taken in the Bale mountains, Ethiopia, at 3,600 metres (© Harald Pauli).

problem, although conservationists might appreciate the reappearance of forest and wildlife.

The desert mountains of the Sahara offered summer pastures to the local nomadic tribes, a land-use system which had existed since prehistoric times. In the deserts of the Andean mountains, settlements were established predominantly in connection with mining.

Establishing biosphere reserves in the Mediterranean life zone seems to be the best strategy for maintaining or restoring cultural mountain landscapes. It seems to be less appropriate to allow abandoned regions to revert to wilderness for which national or regional parks are more appropriate.

Tropical mountains

In tropical mountains, the climate can range from seasonal (Ruwenzoris, Sierra de Santa Martha, Mount Willem, Kinabalu) to slightly seasonal (e.g. Peruvian Andes, Kilimanjaro). Seasonality refers mainly to annual cycles of precipitation. Throughout, however, the diurnal cycle of temperatures exceeds the annual mean. Mountains such as the Ruwenzoris experience 'winter at night, summer during the day'. Beyond altitudes of

4,000 metres, frost occurs every night. The uppermost summits are glaciated. It is important to note, however, that there is no snow season to determine a vegetation period. Related habitats peculiar to temperate mountains, such as snow beds, melt-water fens, avalanche meadows or wind-exposed ridges are absent from these mountains. Here the predominant vegetation types along the elevational gradient are: montane rain forest, cloud forest and giant rosette formations (paramo; note however that giant rosettes do not occur at Kinabalu or in the mountains of Papua New Guinea). There is no elevational limit to vascular plant growth, as indicated by *Poa ruwenzorensis* at the very summits of the Ruwenzoris or Senecio species in the Kibo crater of Kilimanjaro. Owing to the diurnal temperature cycle, soils never warm up. Growth of trees is limited to soil temperatures above +2 degrees Celsius.

Mountains with alternating rainy and dry season (Andes between 14–30°S, and mountains on islands, e.g. Hawai'i) are characteristic for the area close to the Tropics of Cancer or Capricorn. They are dry in the lowlands. On the middle slopes condensation clouds cause a wet environment with cloud forest. The tree line here is not necessarily determined by low temperature but an expression of the dry high-mountain climate. On the other hand, the dry tropical Andes are also known for the highest stands of trees (*Polylepis* species) just above 5,000 metres. Though in some regions giant rosettes still occur (e.g. *Argyroxiphium sandvicense* on Haleakala, Hawai'i) the predominant life form in the alpine zone is tall tussock grass which forms vast expanses of grassland known as 'puna'. Many tropical summits are heavily glaciated. On favourable microsites, however, cryptogams and even vascular plants might grow. From the summit of volcano Socompa (6,000 m, Argentina), around a steam vent, a complete ecosystem was reported, with mosses and lichens as primary producers and small insects, a bird species and a rodent as consumers.

The mountains of the wet tropics have been cultivated up to the level of montane forest (swidden, bananas, coffee etc.). Recent overpopulation and/or introduction of modern agricultural practices (vegetables, ornamentals) have increased pressure on the cloud forests. Significant areas have been cut or burned and transformed into cultivated land. Higher up in the paramo, most regions are still natural; in the case of the Ruwenzoris or the mountains of New Guinea, they might even be considered pristine. This is completely different in the dry seasonal tropics where e.g. the Altiplano in South America hosted the ancient civilisation of the Incas. The Europeans with their livestock interfered with indigenous land use systems, especially with the 'puna' as grazing ground. The native animals (alpaca, guanaco) are adapted to tussock grasses which in particular was not the case with cattle and sheep. Nevertheless, the people in most of the high ground of tropical South America still rely on traditional farming. Most of the biosphere reserves have been established for conservation purposes. Nevertheless, the biosphere reserve strategy would fit well with attempts to save cultural values and to invent sustainable land-use systems benefiting from the tremendous agrodiversity in that region.

The volcano 'Etna' in Sicily, Italy, hosts a unique endemic flora (© Gerhard Hornsteiner, ecoResponse)

Mountains of special interest

Some mountains deserve special attention such as isolated volcanoes, those of islands in particular. Depending on the past, recent or ongoing geological activity, they provide an interesting array of phenomena, such as successions, reaction of biotic communities to recurrent disturbance etc. Island volcanoes such as the Hawaiian volcanoes, Teide on Tenerife, Pico on the Azores, and Etna on Sicily, host a unique endemic flora and fauna, while others may be interesting from a cultural perspective (e.g. Fujisan in Japan, Popocatepetl in Mexico) or attractive for economic purposes (e.g. Iceland making use of thermal energy and tourism, Kamchatka as attraction for tourism). Interestingly, however, not many volcanic regions or single volcanoes have been declared biosphere reserves.

Some of the most spectacular mountains in the world are the Guiana Tepuis in southern Venezuela although they do not have a treeless alpine zone. The reason is that close to the equator, the column-like mountains receive high amounts of rain. The rocky plateaus on the mountain tops are either not vegetated or they have patchy peat which fills rock fissures and hollows. The plants here, exclusively endemic, are adapted to extremely low nutrient levels. Many are carnivorous, although peculiar growth forms such as the tubular *Brocchinias*, have no obvious connection with the habitat conditions.

From a cultural perspective, the many sacred mountains are important. Impressive mountains such as Mount Kailash on the Tibetan Plateau, Fujisan in Japan, Mount Kawagebo in Western China, Mount Olympus in Greece and Mount Sinai are admired by millions of people. These mountains are considered to be the abode of the deities of many of the world's religions and provide an over-arching spirituality and source of myths, legends and psychological balm and aspiration for society at large. Though not sacred in a strict sense, spectacular mountains have attracted the special attention of local people but also of tourists. Among these mountains are the Matterhorn and

Mont Blanc in the Alps, Fitzroy in the Patagonian Andes, the Cordillera Blanca in the tropical Andes, and East African mountains such as Kilimanjaro and Mount Kenya.

Outlook

The mountains of the world, especially the highland areas, are tremendously diverse, which is due to the living conditions in major life zones; differences among the flora and fauna, as a result of evolution and past migrations, the age of orogenesis, and last not least – by climate change and human intervention. Though generally harsh environments for humans, mountains have served as safe havens from enemies, protected people from diseases, provided various resources (or natural pastures, soils for cultivation). They are important for recreation and contribute significantly to the rich biodiversity on the globe. Yet we must also take into account the actual and perceived threats to the lowlands, if mismanagement of mountain resources continues unabated. In this sense, mountains are not only suppliers of many products, they also protect watersheds for the benefit of the lowlands. The converse of 'protection' is that they are potential destroyers of the life-support system of hundreds of millions of people in the plains. If development aid and development are expected to produce immediate returns on investment – then, obviously, it would be rational to neglect the mountains. But such attitudes reflect the short-term view. The long-term implications are becoming increasingly clear. They are inextricably entwined with the growing concern that continuing development must be sustainable. The present book is intended to show that biosphere reserves in mountain regions are long-term instruments for a continuous monitoring and evaluation of natural and anthropogenic changes to mountain ecosystems and to conservation and development of mountain landscapes.

Initially, biosphere reserves served as a concept to maintain long-term research based on the ecosystem approach, when elucidating and explaining the role of humans was the main objective. The modern concept of the Seville Strategy further strengthened this attempt, shifting the focus in favour of local populations. The conservation perspective has declined and will be restricted to reserves of the old generation but not reinvented as a main goal. With the core zone, however, conservation has not been completely abandoned. Overall, the biosphere reserve strategy is flexible, and can be adapted easily to different purposes. Many mountains are facing numerous challenges – now and in the future. Biosphere reserves can be effective tools. Successful examples as presented in this book are the proof.

References

(Books or monographs recommended for further reading have been marked in green.)

Barthlott, W., Lauer, W., Placke, A. (1996). Global distribution of species diversity in vascular plants: towards a world map of phytodiversity. In: Erdkunde 50: 317–327.

Bernbaum, E. (1990). Sacred Mountains of the World, Sierra Club Books, San Francisco.

Breckle, S-W. (2002). Walter's vegetation of the earth. The Ecological Systems of the Geo-Biosphere. Forth Edition, Ulmer Stuttgart.

Burga, K., Klötzli, F., Grabherr, G. (2004). Gebirge der Erde. Ulmer Stuttgart.

Ellenberg, H. (1996). Vegetation Ecology of Central Europe. Cambridge University Press, Cambridge.

Nagy, L., Grabherr, G., Körner, C., Thompson, D.B.A. (2003). Alpine biodiversity in Europe, Ecological Studies 167. Springer, Berlin.

Grabherr, G. (2009). Biodiversity in the high ranges of the Alps: Ethnobotanical and climate change perspectives. In: Global Environment Change 19: 167–172.

Grabherr, G., Mucina, L. (eds.) (1993). Die Pflanzengesellschaften Österreichs, 3 volumes. Gustav Fischer, Jena Stuttgart New York.

Grabherr, G., Gottfried, M., Gruber, A. & Pauli, H. (1995). Patterns and current changes in alpine plant diversity. In: Chapin, F.S. & Körner, C. (eds): Arctic and alpine biodiversity. Ecological Studies 113. Springer, Berlin: 167–182.

Grabherr, G., Gottfried, M., Gruber, A. & Pauli, H. (2010). Climate change impacts in alpine environments. In: Geography Compass 4/8: 1133–1153.

Grötzbach, E. & Stadel, C. (1997). Mountain peoples and cultures. In: Messerli, B. & Ives, J.D. (eds): Mountains of the world.

Höchtl, F., Lehringer, S. & Konold, W. (2005). Kulturlandschaft oder Wildnis in den Alpen? Haupt, Bern.

Jeník, J. (1997). The diversity of mountain life. In: Messerli, B. & Ives, J.D. (eds): Mountains of the World: 199–231.

Johnston, F. & Pickering, C.M. (2001). Alien plants in the Australian Alps. In: Mountain Research and Development 21/3: 284–291.

Körner, C. (2001). Alpine Ecosystems. In: Levin, S.A. (ed.): Encyclopedia of Biodiversity, vol 1. Academic Press, San Diego: 133–144.

Körner, C. (2003). Alpine plant life: functional plant ecology of high mountain ecosystems, 2nd ed. Springer, Berlin. In: Körner, C. & Spehn, E.M. (eds): Mountain biodiversity: a global assessment. Parthenon, New York London: 3–20.

Larcher, W., Kainmüller, C. & Wagner, J. (2010). Survival types of high mountain plants under extreme temperatures. In: Flora 205: 3–18.

Molinillo, M. & Monasterio, M. (2006). Vegetation and grazing patterns in Andean environments: A comparison of pastoral systems in punas and paramos. In: Spehn, E.M., Liberman, M. & Körner, C. (eds): Land use change and mountain biodiversity. Taylor & Francis, Boca Raton.

Messerli, B. & Ives, J.D. (eds) (1997). Mountains of the World – a global priority. Parthenon, New York.

Nagy, L. & Grabherr, G. (2009). Biology of Alpine Habitats. Oxford University Press, Oxford.

Salick, J., Zhendong, F. & Byg, A. (2009). Eastern Himalayan alpine plant ecology, Tibetan ethnobotany, and climate change. In: Global Environmental Change 19: 147–155.

Spehn, E.M., Liberman, M., Körner, C. (eds) (2006). Land use change and mountain biodiversity. Taylor & Francis, Boca Raton.

Stöcklin, J., Bosshard, A., Klaus, G., Rudmann-Maurer, K. & Fischer, M. (2007). Landnutzung und biologische Vielfalt in den Alpen. Nationales Forschungsprogramm 48: Landschaften und Lebensräume der Alpen des Schweiz. Nationalfonds. vdf Hochschulverlag, Zürich.

Tasser, E., Mader, M. & Tappeiner, U. (2003). Effects of land use in alpine grasslands on the probability of landslides. In: Basic and Applied Ecology 4: 271–280.

Troll, C. (1961). Klima und Pflanzenkleid der Erde in dreidimensionaler Sicht. In: Naturwissenschaften 48: 332–348.

Wielgolaski, F.E. (ed): Polar and alpine tundra. Ecosystems of the world 3. Elsevier, Amsterdam.

Wrbka, T., Peterseil, J., Schmitzberger, I. & Stocker-Kiss, A. (2004). Alpine farming in Austria, for nature, culture or economic need? In: Jongmann: The new dimensions of the European landscapes. Springer, Berlin: 165–177.

Mountains are centres of biodiversity: Crocus albiflorus in Gran Paradiso NP, Italy (Nicola Gerard © Archive NP).

Europe's Mountain Biodiversity: Status and Threats

by Marcus Zisenis & Martin F. Price

At the global scale, mountains are centres of biodiversity. For instance, of the 25 global hotspots identified by Conservation International (Mittermeier et al., 2005), all but two are entirely or partly mountainous. Two of these hotspots – the Irano-Anatolian and the Mediterranean Basin – include mountains in southern and southeastern Europe. Similarly, within Europe, most hotspots of plant, bird and mammal diversity are in mountain areas. A number of factors interact to cause these high levels of biodiversity (Körner, 2002). These include the compression of thermal and climatic zones over relatively short distances, steep slopes, the diversity of aspects, variations in geology and soils, and the fragmentation of mountain terrain. In addition, many mountain areas are isolated from one another either in terms of distance or because of unsuitable habitats – at least since the end of the last Ice Age, or because of significant anthropogenic modification of lowland ecosystems – so that species have evolved separately; a major reason for the high levels of endemism in many mountains. Species endemism often increases with altitude (Nagy & Grabherr, 2009; Schmitt, 2009). Within mountain areas themselves, centuries or millennia of human intervention, particularly through burning and grazing, have also been important for maintaining populations of many species and particular habitats in spatially diverse cultural landscapes.

Habitat and species diversity in European mountains

Ranging from the Arctic to the Mediterranean and experiencing climates from the oceanic to the continental, Europe's mountain ecosystems are highly diverse and cover 36 per cent of the continent, including 29 per cent of the European Union (EU) (EEA, 2010b). Across the continent, forests cover 41 per cent of the area of mountain ecosystems and over half of the area of the Carpathians, the mountains of central and south-east Europe, the Alps, and the Pyrenees. As a result of sharp altitudinal gradients in both temperature and precipitation, habitat and species diversity are generally higher in mountain areas than in lowlands (Regato and Salman, 2008). Mountain grasslands, for instance, show remarkable biodiversity, which is comparable to certain types of tropical rainforests (EEA, 2002). To a large extent, this biodiversity derives from centuries of intervention by people and their grazing animals; if grazing or mowing decreases below a certain level, many of these species are lost as plants of higher stature take over (Nagy & Grabherr, 2009). Although alpine areas above the tree line cover only three per cent of Europe's land surface, they host 20 per cent of its native vascular plant species. It is estimated that more than 2,500 species and subspecies of alpine flora are confined to, or predominantly occur, above the tree-line. The proportion of species restricted to the alpine zone varies from less than 0.5 per cent of the total flora in Corsica to about 17 per cent in the Alps (Nagy et al., 2003). Numbers of vascular plants decrease from south to north, whereas numbers of cryptogams (bryophytes and macrolichens) show the opposite trend (Virtanen et al., 2003). Species endemism, in particular, often increases with altitude within mountain regions, partly due to the isolation of populations and speciation processes over geological time scales (Regato & Salman, 2008; Nagy & Grabherr, 2009; Schmitt, 2009). For example, the Caucasus ecoregion has the highest level of endemism in the temperate world, with over 6,500 vascular plant species, at least 25 per cent of which are unique to the region (Wilson, 2006). In the rest of Europe, the highest number of endemics and narrow range taxa are found in the Alps and the Pyrenees, with high numbers also in the Balkan Mountains, Crete and the Sierra Nevada, the Massif Central, Corsica, and the central Apennines (Väre et al., 2003). The mountain regions of the Iberian peninsula (excluding the Pyrenees) show a particularly high number (64) of endemic Species of Community Interest listed in Annexes II and IV of the EU Habitats Directive, followed by the Balkans (24). For individual massifs, the highest number of Species of Community interest is found in the Alps (24 endemic species), followed by the Carpathians (18). The highest number of mountain Species of Community Interest on islands are found on the Canary Islands (30) (ILE SAS-ETC/BD, 2010).

Mountain areas provide important habitats for many bird species. Mountain ranges can also be significant bottlenecks to migration (Heath et al., 2000), which is a key issue as populations of long-distance migrants are 'declining alarmingly' (BirdLife International, 2004); their water bodies and associated wetland

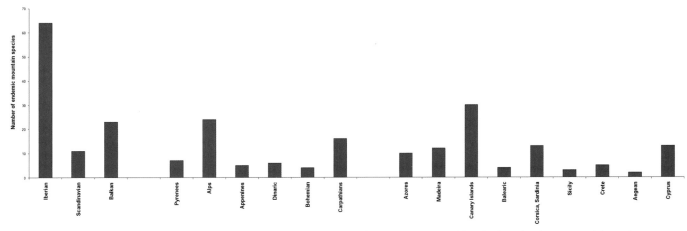

Number of mountain Species of Community Interest (Annex II and IV of the EU Habitats Directive) endemic to mountain regions, mountain ranges, and islands of Europe (Source: ETC/BD, 2008).

grassland ecosystems, in particular, are critical resting sites (EEA, 2010b). Mountain habitats in Europe (mainly forests and agricultural grasslands) are estimated to support 73 priority bird species and contain 558 Important Bird Areas (IBAs). More than half of these species are declining strongly in Europe or even threatened by extinction. Reasons include inappropriate forest management, changes in agricultural practices and poorly planned tourism development (BirdLife International, 2009).

Just as mountain biodiversity varies across Europe, so do human impacts on this biodiversity. Most research has been done on the Alps, but factors such as the density of human activity and its impact on biodiversity differs in from one range to another. This in turn affects policies for the conservation of mountain biodiversity and wilderness: mountain areas are also at the heart of Europe's remaining wilderness areas (EEA, 2010b).

Low-intensity farming supports biodiversity
Low-intensity farming in Europe, particularly livestock rearing and traditional cultivation methods, has created semi-natural habitats that support a range of species such as species-rich grasslands, hay meadows and grazed wetlands. The functional diversity in many ecosystems depends directly on traditional types of agricultural land use and farming practices (Cerquiera et al., 2010). High Nature Value (HNV) farmland is typically associated with low-intensity agriculture, especially grazing. Fifty-one per cent of Europe's HNV farmland is situated in mountain areas (EEA, 2010b).

European mountains support a rich cultural heritage
The specific environmental conditions and resources of mountains – steep slopes, poor and shallow soils, and extreme climate conditions – have also resulted in high cultural diversity and varied adapted land-use practices that reflect traditional knowledge, cultural and spiritual values (Regato & Salman, 2008; Nordregio, 2004). People and nature together form diverse and rich cultures, which attract tourists from the European lowlands and far beyond, supporting a large tourism industry in summer and winter (EEA, 2009).

Threats to Europe's mountain biodiversity
Mountain ecosystems face a complex of rapid changes

Mountain ecosystems are fragile and vulnerable to changes due to their particular and extreme climatic and biogeographic conditions. In the Alps, for example, the main pressures on mountain biodiversity are caused by changes in land use practices, infrastructure development, unsustainable tourism, overexploitation of natural resources, fragmentation of habitats, and climate change (EEA, 2002).

European mountain regions, in general, are experiencing strong climate change (glacier retreat, temperature increases, changes in precipitation), as well as land-use changes due to socioeconomic pressures (EEA, 2009; 2010b). Marginal land in European mountains is being abandoned, while land use is being intensified on productive sites in the lowlands and along the bottoms and lower slopes of mountain valleys (Hagedorn et al., 2010).

Biodiversity suffers from land use intensification and abandonment

In comparison to traditional land-use practices, plant diversity is reduced in the alpine zone by both intensification and land abandonment (Spehn & Körner, 2005). While agricultural management on economically profitable sites in the Alps is being intensified, remote areas or those with potentially lower yields are being abandoned (Kampmann et al., 2008). Mountain grasslands are very vulnerable to decreased use because activities such as regular mowing are important for maintaining high species diversity in certain grasslands (Galvánek & Lepš, 2008). In western Europe, such grasslands are often abandoned in unprofitable locations with steep slopes, poor soils or underdeveloped road infrastructure or where pastureland is infrequently used, becoming overgrown with bushes and trees (Gellrich et al., 2007). A study in the border area between Poland, Slovakia, and Ukraine in the Carpathians revealed similar occurrences in eastern Europe. Here, however, forces such as speculation, unemployment, land-reform strategies and changes in rural population density during the post-socialist period also complicated matters by affecting land ownership patterns (Kuemmerle et al., 2008). Both abandonment and intensified farming of mountainous agricultural land are evident across Europe's mountains. Overall, the area of forest has

Low-intensity farming supports biodiversity in European mountains: impressions from the Bavarian Alps around Mittenwald (© Sigrun Lange).

increased since 1990. At the national scale, changes in agricultural land use have been most marked in the Czech Republic, especially from 1990 to 2000 when the annual rate of land cover change was 1.3 per cent (EEA, 2010b).

Infrastructure development adds to fragmentation

Lowland-focused policies that ignore the vulnerability and disadvantaged character of mountains, and the high demand for mountain resources by lowland people, often worsen human pressures and environmental disturbances in mountains (Regato & Salman, 2008). For example, constructing highways and motorways increases the isolation and fragmentation of mountain natural environments and the number of physical barriers to the natural movement of many organisms (UNEP, 2007). In specific locations, developing skiing infrastructure can cause considerable damage to soils and vegetation. Soils become more vulnerable to water erosion, and hillsides with low vegetation cover have higher water runoff levels, increasing the risk of flooding lower areas. Producing artificial snow increases water consumption, which may disturb the hydrological cycle for habitats of high conservation value such as bogs, fens and wetlands at high altitude (EEA, 2002; 2009).

Unsustainable exploitation threatens ecosystem goods & services

Mountain ecosystems provide diverse goods and services to Europe's population as a whole (EEA, 2010b). However, ensuring the continued delivery of these goods and services requires careful management. One major threat to biodiversity is mass tourism, as development can lead to large-scale damage to nature and landscapes. It also favours the introduction of invasive alien species into native habitats (UNEP, 2007). Invasive species are being encountered at ever higher altitudes (Pauchard et al., 2009; EEA, 2010b). In the Caucasus ecoregion, highly valuable mountain forests are threatened by unsustainable management and exploitation in the form of harvesting wood for fuel and the timber trade. This will lead to irreversible loss of biodiversity and the goods and services on which many local people depend (Williams et al., 2006). Hunting and poaching in the Carpathians generally focus on rare and endangered species such as large carnivores, eagles, owls, chamois, marmots and many small invertebrates and plants. As their populations are small and isolated, they may not maintain long-term viability and become extinct (UNEP, 2007).

Severe consequences of climate change

Climate change threatens important mountain ecosystem services, including supporting rich biodiversity heritage and providing freshwater to vast lowland areas (EEA, 2010b). Climate change is affecting Europe's mountains in different ways. At the regional level, changes in temperature and precipitation result in changes in snow cover, glacier volume and extent, permafrost and surface runoff (EEA, 2009). In the Alps, average temperatures increased by approximately two degrees Celsius between the late 19th and early 21st centuries. This was more than twice the rate of change in the Northern hemisphere as a whole (Auer et al., 2007) and resulted in significant loss of glacial volume (e.g. Zemp et al., 2007). The rising temperature will increase the proportion of precipitation falling as rain instead of snow, so that there will be more runoff in winter and less in spring and summer (EEA, 2009). Changes in precipitation in the Alps have already been associated with changes in vegetation (Cannone et al., 2007). The frequency of natural hazards such as mudflows, floods and droughts is expected to increase. Climate change also affects many mountain ecosystems directly and indirectly together with other factors such as economic and planning policies (Price, 2008). The sensitivity of mountain biodiversity to climate change has been shown by models and validated by in situ observations of phenomena such as upward shifts of vascular plants and changes in species composition at Mount Schrankogel in the Austrian Alps (Pauli et al., 2007). There are projections that the tree line could shift upward by several hundred meters (EEA, 2009), and evidence that this process has begun in Scandinavia, the Urals, the western Carpathians and the Mediterranean (EEA, 2010b).

Flora and fauna are expected to migrate upwards in order to stay within their bioclimatic envelope. Evidently, however, there is no upward escape from the top of a mountain. Sixty per cent of mountain plant species in the Alps may face extinction by 2100 if they cannot adapt to climate change by moving northwards or upslope (EEA, 2009). Many alpine species have limited dispersal capabilities (Nagy & Grabherr, 2009), and habitat fragmentation may further limit their mobility (Higgins et al., 2003). Small isolated populations face bottlenecks, which decrease their genetic viability and adaptability to a changing environment and may cause extinction in the long term. Species and habitats associated with water bodies, flowing water, and wetlands are likely to be especially affected by the expected shifts in water regimes. These include less precipitation and runoff in summer and more in winter, runoff peaks earlier in the season, a shorter duration of snow cover and melting of glaciers and permafrost. A temporary habitat enlargement can be foreseen for some macrofauna in the Alps, for instance the ibex (*Capra ibex*), the Alpine chough (*Pyrrhocorax graculus*), and the rock partridge (*Alectoris graeca*). Other more isolated species populations such as snow finch (*Montifringilla nivalis*), water pipit (*Athus spinoletta*) and ptarmigan (*Lagopus mutus*) are threatened by global warming (Niedermair et al., 2007).

Addressing the challenges

Mountain regions in Europe vary in terms not only of their biogeographic environmental conditions but also their political and socio-economic circumstances (EEA, 2009; Nordregio, 2004). Equally, our knowledge of these very diverse environments varies greatly with, in particular, much more knowledge regarding the Alps than other regions (EEA, 2010b). Nevertheless, it is notable that, across Europe as a whole, particularly large pro-portions of the mountains of almost all countries are designated as protected area. Within the EU, 43 per cent of the total area of Natura 2000 sites is in mountain areas; 92 per cent of the mountain area has been designated as Less Favoured Areas (LFAs) and 17 per cent as HNV farmland (EEA, 2010b).

European and international legal frameworks can serve as tools to mitigate severe pressures such as climate change through targets and actions to reduce greenhouse gas emission reductions agreed at global (UNFCCC, Kyoto Protocol) and EU levels, and to adapt to some inevitable climate change. However, there are many complex interacting reasons for negative trends in biodiversity, which are often driven by national forces (e.g. employment and income imbalances), European activities (e.g. Common Agricultural Policy) and even global policies. This implies a need to build on existing regional initiatives such as the Alpine Convention and the Carpathian Convention (EEA, 2010b), to foster other trans-national initiatives, and to integrate management strategies, which should be developed and implemented with the active participation of the public concerned and the relevant stakeholders (Partidário et al., 2009; Fonderflick et al., 2010).

Measures to increase ecological connectivity are particularly important, especially within and between the many mountain ranges along national borders (Worboys et al., 2010). As for each major ecosystem type in Europe, it is essential to monitor the success of regional mountain biodiversity actions and to undertake applied research (Borsdorf & Braun, 2008) and targeted public relations (UN, 1992; CBD, 2010; GMBA 2010).

Acknowledgements

This chapter is excerpted mainly from '10 messages for 2010: Mountain ecosystems' (EEA, 2010a) and also from EEA (2010b). The message was written by Marcus Zisenis (ECNC, ETC/BD), with contributions from Dominique Richard (MNHN, ETC/BD), Martin Price (Perth College UHI), Amor Torre-Marin and Lawrence Jones-Walters (ECNC), Luboš Halada, Peter Gajdoš, and Július Oszlányi (ILE SAS). Furthermore, the authors would like to acknowledge valuable comments from Branislav Olah and André Jol (EEA), Marco Fritz (DG Environment of the European Commission), the EIONET National Reference Centres (NRCs), Sabine McCallum (ETC/ACC), and Ivone Pereira Martins (EEA); and the EEA project manager of the '10 messages for 2010', Frederik Schutyser.

References

(For further references cited in the text, please refer to the '10 messages for 2010' at: http://www.eea.europa.eu/publications/10-messages-for-2010-mountain-ecosystems)

BirdLife International (2004). Birds in the European Union: a status assessment. BirdLife International, Wageningen.

EEA (2010a). 10 messages for 2010: Mountain ecosystems. EEA, Copenhagen.

EEA (2010b). Europe's ecological backbone: recognising the true value of our mountains. EEA, Copenhagen.

ETC/BD (2008). Habitats Directive Article 17 Report (2001–2006). Available at: http://biodiversity.eionet.europa.eu/article17.

Heath, M.F.; Evans, M.I.; Hoccom, D.G.; Payne, A.J. and Peet, N.B. (eds.) (2000). Important Bird Areas in Europe: Priority Sites for Conservation. Volume 1: Northern Europe, Volume 2: Southern Europe. BirdLife International, Cambridge.

Mittermeier, R.A.; Robles, G.P.; Hoffman, M.; Pilgrim, J.; Brooks, T.; Goettsch Mittermeier, C.; Lamoureux, J. & da Fonseca, G.A.B. (2005). Hotspots Revisited: Earth's Biologically Richest and Most Endangered Ecoregions. University of Chicago Press, Chicago.

Mountain landscape in Cajas National Park, Ecuador (© Sigrun Lange).

Sustainability and the Biosphere Reserve: A Compromise between Biodiversity, Conservation and Farmscape Transformation

by Fausto O. Sarmiento

Two important concepts collide when analysing sustainable development and biodiversity conservation in mountain biosphere reserves. Firstly, the actual meaning of both, sustainability equated with maintenance, and development equated with growth, have to be grappled with in the context of the poorly known mountain ecosystems operating in limited, and often ineffective, economic trends in mountain communities. Secondly, the actual meaning of conservation of nature and natural resources, equated with the maintenance of biological diversity found in the wild lands, isolated and often mature wilderness of mountainous regions. Finally, in order to better understand the intricate mesh of transactions between the mountain environment and the cultural landscape of mountain-dwelling people, we have to grapple with a science/art management dilemma: While most conservation practitioners use these seemingly easy terms now almost interchangeably, special clarification of these terms is needed from scholarly scientific and artistic sources (Webersik & Wilson 2009, Coelho et al. 2010).

Because of the difficulty of defining sustainable development and its many implications, the term is considered a constructed mantra (Van der Ryn & Cowen 1995, Bejan & Lorente 2010) as every profession uses it to describe some fine-tuned concept of permanency in time, improvement in space, ascendancy in effects and security for intergenerational equity. These are required elements of a sustainable system, but difficult to come to terms with in the practical sphere. Defining mountain-sustainable development brings another dimension that will be analysed below by means of a discourse analysis. Since everybody had a different definition, Hamilton (1996) proposed to utilise the inverse-definition approach to better grasp sustainability meanings. No unique definition exists for sustainability in a lifetime frame and no example is given of something sustainable; however, a clear and immediate comprehension of unsustainable practices can be found in a deforested watershed, in an eroded slope, in a polluted brook or in an acculturated mountain village. They are examples of what sustainability is NOT and, hence, sustainability ideas are readily understood.

Biological diversity also has its derivation, but most scholars agree that there are at least three levels of biodiversity research and management: genetic, population and ecosystem. It is obvious that flora and fauna are mostly referred to by the media in relation to biodiversity conservation; however, in the mountain landscape there are many organisms that do not fit into those two mega-categories, such as bacteria, virus, fungi, lichens and protobios such as spores, pollen, cysts and other latent life. Moreover, as mountains are often cultural landscapes where the relation between humans and environment has endured through millennia, many elements of the biodiversity complex are produced by the action of people. Domestication of wild plants and animals has produced complex organisms that are now living depending on the dispersal role of humans associated with the plant or animal, so that agro-biodiversity is an important indicator of conservation efforts (Lewis & Chambers 2010). It is precisely these cultural landscapes that receive priority under the scheme defined by the biosphere reserve approach, whereby an inner nucleus of mature wilderness can be found surrounded by a buffer area with less anthropogenic disturbance than the surrounding transitional area with a fully developed farmscape. The three-layered approach of biosphere reserve conservation strategy (core zone, buffer zone and transition zone in the periphery) resembles the theoretical construct by geographers dealing with the central-place theory, indirectly prioritising 'pristine', 'untouched' or 'virgin' forest compared to managed landscape features. It provides a workable strategy to help conservation areas – mainly big national parks, ecological reserves and wildlife sanctuaries – balance biological and cultural diversity with sustainable economic development of the surroundings.

By having the network of biosphere reserves, coordinated by the United Nations' Education, Science and Culture Organisation (UNESCO) the system is used to test, refine, demonstrate and implement projects often conflicting with nature conservation, economic development and cultural values (Schaaf 2006). There are some 109 countries that have accepted and initiated biosphere reserve conservation; the majority of them are mountainous areas (MAB 2010). The structure of protection within the World Network of Biosphere Reserves has 564 units, and in the United States of America, out of 47 units, only 29 are managed by the National Parks Service. Compare this to Ecuador, where its four units are all state or federally controlled and managed. Whether islands or highlands, the vertical dimension is one of the most important conditions for long-lasting livelihood security, and thus, a leading driver in sustainability research and applications.

Mountain cognition and onomastics

The same controversy is generated by the term Mountain. As Debarbieux & Gillet (2000) pointed out, the concept of mountain is challenged according to the region of the world, the cultural background, even the disciplinary affiliation. There is no agreement on whether montology, the science of mountain studies, exists or should exist (Rhoades 2007), but the fact is that in many countries the confusion lends itself to differential appropriations of the common good in the highlands (Debarbieux & Price 2008) or the actual cognitive process of nature and its resources (Sarmiento 2001a). Many of the cognitive processes include different technical and biophysical characteristics, while others come from a socio-cultural background. There are discrepancies in the calculation of the mountain mass, depending on calculating its total edifice or only its prominence or autonomous height. The same discrepancies arise when looking at the distinction of being the tallest mountain on earth: If you follow convention, the elevation above sea level is taken as the most common parameter, making Mt. Sagarmantha (previously called Everest) the tallest on Earth. If you calculate the altitude by measuring the radius from the centre of the planet instead, Mt. Chimborazu in Ecuador is the tallest, owing to a longer equatorial distance of the oblate spheroid planetary geometry. In addition, if you consider the length of a continuous slope as the determinant of the mountain, the tallest would be Mt. Mauna Loa in Hawai'i, arising from the bottom of the sea and climbing to the summit in one long slope. Furthermore, if you believe that verticality relates more to the proportion of flat terrain surrounding the elevated point, Mt Kosciusko in Australia would be the tallest on Earth, with Mt. Kilimanjaro a close competitor. Lastly, if the mountain is defined as the general edifice, Mt. Lam Lam in the Pacific Islands would be the tallest on Earth, arising from the depths of the Marianna's trench and protruding to open air in Guam. Another important difference relates to place naming, or onomastics, based on the description given by the words used to describe mountains. Several places retain the vernacular nomenclature, such as the Himal region in the Himalayas. Other mountains have lost their vernacular descriptors to favour either Latin names or the nomenclature of the empire, being English, French, Portuguese, Spanish and so forth. This is exemplified by the term 'Andes' formerly thought to be the name ascribed to a bellicose tribe of Peru, now confirmed rather as the derivation of the name in Castilian shorthand describing the 'andenes' or terraces and 'andenerías' or terrace systems, ubiquitous in the mountains of the Americas at the time of the Spanish conquest. Recently, in many countries, the revival of indigenous identity has obtained controversial renaming of the mountains with the vernacular name sanctioned by law. Ayers Rock in Australia no longer receives this old name; now, most call it Mt. Uluru. Mt. McKinley in the United States of America is now called Mt. Denali. Even in scientific circles, the use of 'alpine' to describe highland grasslands, such as in the Alps of Europe, is being cautioned and replaced with the regional appellations, such as 'Andean' or 'Afro-alpine'.

Hence, the way we conceive mountains depends upon our collective appropriation of historical and socioeconomic power relations and this could affect the way biodiversity conservation is discussed around the world. In some areas of the world, the presence of mountains is associated with inhospitable places, corners of existence where stresses of life are found in rare, often endemic species, such as in Greenland, the Antarctic or even New Zealand and Scandinavia. In some other areas of the world, the presence of mountains is more associated with the familiar place of abode, the domesticated areas where ancestral wisdom has developed its unique livelihood, such as in Indonesia, Malaysia, China, India, Guatemala, Colombia, Ecuador, Ethiopia, Kenya, South Africa and many other countries. The need to revise the cognition and onomastics of the mountains to understand the priority given to conserve them for future generations exists: on the one hand, the untouched wilderness, and on the other, the man-aged cultural landscapes offer a continuum of choices to manage the areas for biodiversity conservation in any one category or a combined approach of conservation categories accorded by the World Conservation Union (Dudley 2009, Jenkins & Joppa 2010). Biosphere reserves have been designed to serve as the meeting ground where both objectives can be tested, experimented, demonstrated and implemented.

Mountains as ecoregions of global importance

With the onset of globalisation and the practice of modern approaches to biodiversity conservation, such as transboundary protected areas, connectivity conservation – landscape ecology, macroecology and the like – new techniques have appeared from hybrid disciplines to make scientific principles applicable in remote areas with computerised technologies (Price et al 2004), generalised in all countries, making plausible global programmes and worldwide networks viable. Mountains are no longer isolated, marginal, excluded and forgotten places; rather, they are of central importance for water catchments and distribution, military and strategic operations, communication hobs with antennae and repeater stations, observatories and pilot communities for alternative development scenarios, most of them threatened by global change (Sarmiento 2009). Many of such programmes have successfully developed baseline information on biodiversity capital and vulnerability (Körner & Spehn 2002), as well as other mountain-oriented networks that increased our understanding on physical and human geographies, such as glacial retreat (Grabherr et al 2010), climate change (Becker & Bugmann 2001), mountain protected areas (WCPA 2010) and others.

As a new trend in regional geography, mountains are becoming a favourite theme for cross-listing several disciplines working on disparate topics but in the same ecoregional framework or montology. In the past, regional geography mainly concentrated on understanding continent-wide aspects of physical and cultural geography, learning about Africa or Latin America as a subject matter. At present, conversely, regional geography emphasises macroecological processes that appear in the ecoregions of the world, independent of their continental location, because the same phenomena typical of mountains can be found on each and every continent. Therefore, mountains are good examples of ecoregional emphasis that comprises both the biota of the orobiome, as well as cultures/nations of the anthrome, both elements that depend on the verticality of terrain, as exemplified in tropical mountains (Menhard & Sarmiento 2010). Yet, demographics and social constraints in mountain communities continue to be the main driver that hinders sustainability and challenges biodiversity conservation (Sarmiento 2001b).

Sustainable development in the context of BRs

The ample gamut of choices for managing biodiversity conservation and the diverse options for economic development of rural people living in or around the protected area demanded an inclusive, hybrid, heterodox and flexible approach to achieve sustainability. The creation of national conservation systems with different categories of conservation territories and the charge of biodiversity to the public trust is rather a new phenomenon. Despite some conservation areas created at the dawn of the 20th century, the majority of them were established in earnest during the 1970s and 80s, in an effort coordinated by some of the United Nations organisations, including FAO, UNESCO, UNEP and UNDP and many donor organisations for foreign aid from the North, mainly state-funded (e.g. CIDA, US-AID, CODESU, GTZ, FCD) and private NGOs as well as foundations (e.g. Nature Conservancy, WWF, Conservation International, the Mellon and the MacArthur Foundation).

With funding support and international conventions to protect biodiversity, many areas of countries that were unoccupied, or sparsely populated, were immediately declared protected areas with the only purpose of nature conservation. The fact that some of the areas have been used by humans since prehistoric times did nothing to quash the notion of 'pristine fauna and flora' that could be protected within the newly declared territory. A good example is the mountains of southern Ecuador, near the city of Cuenca, where pre-Columbian occupation is evident in Paredones and other páramo areas of the El Cajas National Park, that was before considered a National Recreation Area for the abundant fishing of introduced trout available to the citizens of Cuenca and weekend visitors. The new status as National Park puts the páramo as the ecosystem to be conserved as if it were 'natural' in the El Cajas mountains. However, the fact that the anthropogenic grassland existed there amidst other evidence of human agency, was hidden in the conservation agenda favouring the new 'environmental services' function of water catchments over the social/cultural services that this ancestral territory offers.

At present, many countries have declared their national conservation systems with the potential of establishing other supra-national categories available; namely, World Heritage Site and Biosphere Reserves. In some extreme cases, the priority of conservation is such that many of these categories coalesce into a tiered administration whereby different instances respond to different officers and distinct policies on the same territory. Take the Galapagos Islands in Ecuador, which not only is one of the best-known national parks in the world, but also a Marine Reserve, a Special Insular Territory, a Military Outpost Reserve, a Biosphere Reserve and a World Natural Heritage Site, all at the same time.

The Galapagos Islands in Ecuador are nominated a national park, a marine reserve, a special insular territory, a military outpost reserve, a world natural heritage site and a biosphere reserve, all at the same time (© Sigrun Lange).

Biodiversity conservation and sustainability

One of the best alternatives to accommodate sustainable economic development with biodiversity conservation is the Biosphere Reserve model. Here, by segmenting the area into zones of intense, intermediate and absent human activity, priority for biodiversity conservation is given on the assumption that the biota will respond in the same way humans do to the central-place theory. Important concentrations of wild species of plants and animals are thought to occur only in the core zone, mostly excluded from intervention. The flexibility of management in the buffer zone allows mostly 'slash-and-burn' agriculture and low-impact extraction of leaves, roots, fruits, and other non-timber forest resources, so that the physical structure of the forest is maintained with minimal change resulting from shifting cultivation. In the peripheral Transition Zone, marginal settlement for subsistence agriculture allows for small patches of productive parcels amidst the fringes of the forest or other marginal presence of biodiversity, sometimes organised in agricultural cooperatives and affiliated to minimal distribution networks and basic urban structures, such as villages and rural settlements. The pressure of development, however, puts human-dominated landscapes into perspective where you could have a rural village as core, the milpas/chacras parcels as the buffer zone and the mature forest in the surrounding transition area, such is the case in the Quijos river valley of Ecuador (Sarmiento 2008). Rurality plays an important role in the establishment of the biosphere reserve unit, as the underlying theme of the protected area becomes meaningful to small groups of people using traditional techniques, without the emphasis on commodity exports to urban markets, but rather, subsistence lifestyle maintained over generations. This trend may no longer be valid in today's farmscape transformation of tropical countries (Gordon & Sarmiento 2010) owing to amenity migration and other appropriations of the global North.

The problem, as it were, manifests when the supposed actors of the central-place theory do not conform to the norm. This is the case in migrant species, or large foragers and vagrant populations which require large tracts of land or even a home range of massive distances, whereby the connectivity of isolated national parks and biosphere reserves tends to recreate original travel routes within ecological corridors. There are species present only in areas that received human imprint: Even inside the Amazonian forests, the Chonta palm (*Bactris gasipaes*) remains as an indicator of the failure of the pristine myth. On the other hand, some species survive only in plazas or parks in the mountains where their wild relatives have disappeared, such as the Quito coconut palm (*Parahubea cocoides*); or were never present, as the cultivated palm remains as testimony to pre-Columbian southern migrations. Another example of the mountains is the Andean wax palm (*Ceroxylum alpinum*) that remains only in isolated patches of cloud forest amidst the anthropogenic grasslands in the highlands. An intriguing research question for further exploration remains open with the inverse Biosphere Reserve model described for the Quijos river valley and the notion that there could be a rural core if we assumed the principle of best management practices for protected landscapes (Brown et al 2005.)

Livelihood protection as conservation wave

As new waves of conservation have come and gone, priorities for biodiversity conservation remain subtle and reminiscent of scientific hegemony and politics of translation and acculturation. This process has not been structured and has not appeared in segments defined by time or in policies defined by funding sources. At the beginning, the first wave of conservation came with the emphasis of the species level. Endangered species were at the crux of the management and illegal export of animals and plants, and the pet trade was the target strategy. Later, the second wave of conservation came with the emphasis on eco-systems, whereby the disappearance of expanses of forest imperiled the natural habitat. A third wave of conservation emphasised the taxonomic importance of rare and endemic species, whereby biodiverse complexity became the priority; no longer the species or the habitat were the main focus, but the collective diversity of biota. A fourth wave of conservation emphasised the required condition of maintenance and improvement of the diversity scenario in the long term, with the realisation that, for successful implementation, humans have to be taken into account for successful implementation. This sustainability wave impacts on the people who live and work in and around the protected areas to become active participants as custodians of the natural capital entrusting them to keep it as it is for the long term. A fifth and final wave of conservation deals with the need to incorporate the innate tendency of growth and human development, both economically and socio-culturally. The poverty alleviation wave aims to address the livelihood (in)security felt in areas that still harbour significant biodiversity value. Rich communities protect their resources better than poorer ones; hence, if poor people progress to a better economic outlook in the developing world, they could also become agents of protection of biodiversity, making them stakeholders in the conservation programme and integrating the goals of management with sustainable economic development and social/cultural invigoration.

Conclusion

The International Network of Biosphere Reserves has the distinctive ability to work with different conservation waves at the same time; however, the call for action comes after the most recent emphases on biodiversity, sustainability, and livelihood protection or poverty alleviation. Successful models have been developed and implemented by Biosphere Reserves within the three recent waves. After all, the construction of biodiversity conservation within a sustainable economic development is possible, as conservation is not only science, but also art.

References

Becker, A. & Bugmann, H. (eds.) (2001). Global Change and Mountain Regions: The Mountain Research Initiative. IHDP Report 13, GTOS Report 28 and IGBP Report 49.

Bejan, A. & Lorente, S. (2010). The Constructal Law of Design and Evolution in Nature. Philosophical Transactions of the Royal Society B. 365: 1335–1347.

Brown, J., Mitchel, N. & Beresford, M. (eds.) (2005). The Protected Landscape Approach: Linking Nature, Culture and Community. World Conservation Union IUCN: Gland and Cambridge.

Coelho, P., Mascarenhas, A., Vaz, P. Dores, A. & Ramos, T.B. (2010). A framework for regional sustainability assessment: developing indicators for a Portuguese region. Mountain Research and Development 18(4): 211–219.

Debarbieux, B. & Price, M. (2008). Representing Mountains: From Local to Global Common Good. Geopolitics 13(1): 148–168.

Debarbieux, B. & Gillet, F. (eds) (2000). Mountain Regions: A Research Subject? Brussels: European Commission.

Dudley, N. (2009). Guidelines for Applying Protected Area Management Categories. World Conservation Union: Gland.

FAO (2004). The Challenge of Sustainable Mountain Development. FAO Newsroom. United Nations' Food and Agriculture Organisation. Rome.

Gordon, B., Sarmiento, F., Jones, J. & Russo, R. (2010). Sustainability Education in Practice: Appropriation of Rurality by the Global Migrants of Costa Rica. Journal of Sustainability Education 1(1).

Grabherr, G., Gottfried, M. & Pauli, H. (2000). GLORIA: A Global Observation Research Initiative in Alpine Environments. Mountain Research and Development 20(2): 190–192.

Hamilton. L.S. (1996). The Role of Protected Areas in Sustainable Mountain Development. Parks 6(1): 2–13.

Jenkins, C.N. & Joppa, L. (2010). Considering protected area categories in conservation analyses. Biological Conservation 143(1): 7–8.

Körner, Ch. & Spehn, E.M. (2002). Mountain Biodiversity: A Global Assessment. Parthenon Publishing: London.

Lewis, L.R. & Chambers, K.J. (2010). Introduction: Geographic Contributions to Agrobiodiversity Conservation. The Professional Geographer 62(3): 303–304.

MAB, UNESCO Man and the Biosphere Programme (2010). MAB International Network of Biosphere Reserves. (http://portal.unesco.org/science/en/ev.php-URL_ID=4801&URL_DO=DO_TOPIC&URL_SECTION=201.html)

Menhard, D. & Sarmiento, F. (2010). Vulnerability of tropical mountain communities to global change: the case of Honduras. Journal of Sustainability Education 1(1).

Price, M., Jansky, L. & Iatsenia, I. (eds.) (2004). Key Issues for Mountain Areas. University of the United Nations: Tokyo.

Rhoades, R. (2007). Listening to the Mountains. Kendall/Hunt Publishing Company: Dubuque, Iowa. 184pp.

Sarmiento, F.O (2001a). Les enjeux de la recherche sur les montagnes en matière de terminologie et de connaissances: application à l'espace andin. Revue de Géographie Alpine 89(2): 73–77.

Sarmiento, F.O. (2001b). Mountain Regions: Sustained livelihood for an increasing population? Schwerpunkt: Berge. Entwicklung Ländlicher Raum 6: 16–18.

Sarmiento, F.O. (2008). Agrobiodiversity in the farmscapes of the Quijos River in the Tropical Andes, Ecuador. Pp 22–30. In: Amend, T, J. Brown, A. Kothari, A. Phillips and S. Stolton. (Editors). Protected Landscapes and Agrobiodiversity Values. Volume 1 in the series, Protected Landscapes and Seascapes. IUCN & GTZ. Kaspareg Verlag, Heidelberg.

Sarmiento, F.O. (2009). Darkening Peaks: Glacier Retreat, Science and Society. Book Review. Ecological and Environmental Anthropology 5(1). http://eea.anthro.uga.edu/index.php/eea/article/viewFile/84/61

Schaaf, T. (2006). UNESCO's Role in the Conservation of Mountain Resources and Sustainable Development. Global Environmental Research 10(1): 117–123.

Van de Ryn, T. & Cowen, S. (1995). Ecological Design. Island Press: Washington DC.

Webersik, C. & Wilson, C. (2009). Achieving environmental sustainability and growth in Africa: the role of science, technology and innovation. Mountain Research and Development 17(6): 400–413.

World Commission of Protected Areas (2010). IUCN-WCPA Mountains Biome. Mountain Protected Areas Network. (http://protectmountains.org/)

Almost 17 per cent of the total mountain area is under protection (Enzo Massa Micon © Archive Gran Paradiso NP).

Towards Effective Conservation in Mountains: Protected Areas and Biosphere Reserves

by David Rodríguez-Rodríguez & Bastian Bomhard

People and mountains

Mankind has always had a special relationship with mountains. Mountains have been historically considered as spiritual places, remote wilderness, beautiful landscapes or sites for natural resource extraction. More recently, the focus has been on mountains as important places for the provision of a variety of services such as outdoor recreation, water provision, income generation for local populations, conservation of traditional culture, biological and agricultural diversity and ecological processes (EEA, 2010a; Macchi, 2010; UNEP-WCMC, 2002). People's particular attraction towards mountains as well as low levels of economic exploitation and human habitation in mountains explain why most of the early protected areas were declared in mountain environments on the grounds of 'virgin' nature and breathtaking landscapes and why a substantial proportion of the world's protected area – almost a third outside Antarctica – is still concentrated in mountain areas (Kollmair et al., 2005). Recognising that defining mountains has always been a challenge, here we follow a definition developed by Kapos et al. (2000) which is based on a combination of two main factors: elevation and slope. Six mountain classes were originally categorised under this definition, and a seventh class was included in the 2002 revision of this classification system. The seven mountain classes are shown in Tab. 1 (UNEP-WCMC, 2002). Following this widely accepted definition, mountain areas cover over 39 million square kilometres world-wide, representing 26.5 percent of the global terrestrial surface.

Protected areas in mountain environments

Mountain areas harbour a great deal of biodiversity, both in terms of richness and endemicity (Macchi, 2010). As a result, half of the global biodiversity hotspots are in mountain regions (Kohler & Maselli, 2009). However, mountain biodiversity is also fragile. Many mountain ecosystems such as glaciers, alpine meadows and scrub, as well as mountain freshwater habitats are among the ecosystems most affected by climate change (Kohler & Maselli, 2009; Macchi, 2010). Some of their constituent plant and animal communities face serious extinction risks as a result of reduction in their distribution ranges owing to shifting climatic conditions.

Mountain Class	Criteria
1	Elevation > 4,500 metres
2	Elevation 3,500 – 4,500 metres
3	Elevation 2,500 – 3,500 metres
4	Elevation 1,500 – 2,500 metres and slope ±2°
5	Elevation 1,000 – 1,500 metres and slope ±5° or local elevation range (7 km radius) >300 metres
6	Elevation 300 – 1,000 metres and local elevation range (7 km radius) >300 metres
7	Isolated inner basins and plateaus less than 25 km² in extent, which, although surrounded by mountains, do not themselves meet criteria 1 – 6

Tab. 1: Mountain classes (Source: UNEP-WCMC, 2002).

Changing climatic conditions may also result in species colonising higher altitudes and competing with or substituting original biotas following the rise in temperature (UNEP-WCMC, 2002).

Not only mountain biodiversity is threatened by global change. A number of other important services provided by mountain areas, both tangible and intangible, are also at risk. These include water provision, carbon sequestration, supply of natural resources, cultural diversity, aesthetic landscapes, recreational and spiritual values (EEA, 2010a; Macchi, 2010; UNEP-WCMC, 2002). The speed with which these changes are being felt in mountain areas calls for immediate action to mitigate their effects and, where possible, to facilitate adaptation (EEA, 2010a; Kohler & Maselli, 2009; Macchi, 2010).

Since the first modern protected area (PA), mountainous Yellowstone National Park, was declared back in 1872, the area protected in mountain environments has increased continuously, especially between the 1940s and 2000 when it grew exponentially (Kollmair et al., 2005).

We used the January 2010 version of the World Database on Protected Areas (WDPA; IUCN & UNEP, 2010) to calculate how much of the world's mountain area was protected by the end of 2009. We considered only nationally designated protected areas with known extent. Where the WDPA did not include boundary polygons, we created circular buffers based on the relevant point location and known extent of the corresponding protected areas. Our results are approximate only because the buffered points may neither be a precise representation of the shape of these protected areas nor, thus, show their real overlap with mountain areas and other protected areas. Overall, mountain protected areas cover more than 5.6 million square kilometres worldwide. This represents over 16.5 per cent of the total mountain area outside Antarctica (Fig. 1). Historically, mountain protected areas have made up a large amount of the total area protected. Even today, as much as 32.5 per cent of the entire terrestrial protected area excluding Antarctica lies within mountains (Tab. 2).

Biosphere Reserves in mountains

Biosphere Reserves (BRs) are areas of terrestrial and coastal or marine ecosystems which are internationally recognised under the UNESCO Man and the Biosphere (MAB) Programme on the grounds of biodiversity conservation, sustainable economic development and environmental research and education (UNESCO, 1996). BRs are made up of three zones: one or more core zones, devoted mainly to biodiversity conservation and research; a buffer zone, usually surrounding core areas, where sustainable, low-impact economic, research and education activities are promoted; and a transition zone, where wider land uses are allowed to achieve sustainable and cooperative development. 'Biosphere Reserve' is not a formal category of protected areas, though. Some countries have adopted the denomination 'BR' as one of their categories of protected areas under national legislation. However, these 'BRs' may or may not coincide with UNESCO's internationally designated BRs. In other countries, BRs overlap completely or partially with different categories of nationally or internationally designated protected areas, such as National Parks, World Heritage Sites or Wetlands of International Importance (UNESCO, 1996).

We analysed the area covered by UNESCO's BRs worldwide and compared it to the area they cover in mountain regions to see how well these regions are represented in the BR Network

(Fig. 2). For this purpose, we again used the World Database on Protected Areas (IUCN & UNEP, 2010) which, despite its deficiencies in terms of complete and accurate information for BRs, still currently provides the best available spatial data on a global scale. The BR layer in the WDPA includes only points. Therefore, we buffered each point according to the reported area of the corresponding BR, potentially resulting in additional inaccuracies similar to those noted above for point PAs.

The most recent version of the WDPA includes a total of 514 BRs, covering some 4.9 million square kilometres, i.e. 91 per cent of all the 564 internationally recognised BRs (UNESCO, 2010). We found that 322 (62.6%) of the 514 BRs in the WDPA are located in mountain areas, accounting for over 1.2 million square kilometres or 29 per cent of the total land area covered by BRs. According to these figures, mountain areas are well represented within the BR Network, accounting for almost two thirds of all UNESCO-designated BRs in the WDPA and for well over one quarter of the whole area covered by BRs. Together, the 322 BRs in the WDPA cover 3.7 percent of the world's mountain area (Table 2).

BRs in mountains are not evenly distributed across different regions (Fig. 2), as there are large regions where mountain BRs are absent or scarce in numbers or extent: Scandinavia, North-East Russia, the Balkans, the Middle East, Arab Peninsula, Central and East Asia, Oceania, Central and Southern Africa, and North America. In contrast, mountain areas in South America, Greenland, North Africa and Central and Western Europe are relatively well represented within the BR Network.

Outlook

Although mountain areas are currently well represented within PAs and BRs (Tab. 2; see also Kollmair et al., 2005; Kohler & Maselli, 2009), a targeted expansion in protection both in terms of number of sites and area is still desirable, especially in regions which are currently characterised by high biodiversity and low levels of mountain protection. This strategy would contribute to enhancing the resistance and resilience of mountain ecosystems by means of better connectivity and more sustainable land use (EEA, 2010b).

Nevertheless, even if further protection be implemented, effective protection of mountain biodiversity and the variety

	Total area (x 10³ km²)	Percentage of world's land area	Percentage of world's mountain area	Percentage of world's land PAs or BRs
Mountains (incl. Antarctica)	39,330	26.5	100	
PAs in mountains	5,610	4.2	16.9	32.5
BRs in mountains	1,226	0.9	3.7	28.9
PAs and BRs in mountains	6,148	4.6	18.5	31.7

Table 2: Summary data on mountain area covered by Protected Areas (PAs) and Biosphere Reserves (BRs), overlaps excluded. Unless otherwise noted, Antarctica, which is covered by an international treaty, has been excluded from all calculations on protected area coverage.

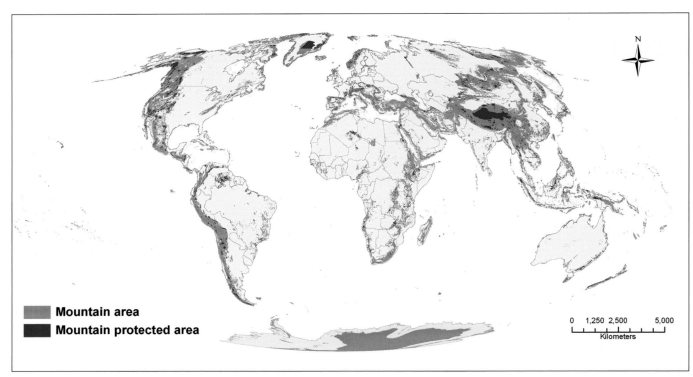

Fig. 1: Distribution of mountain areas and mountain protected areas worldwide.

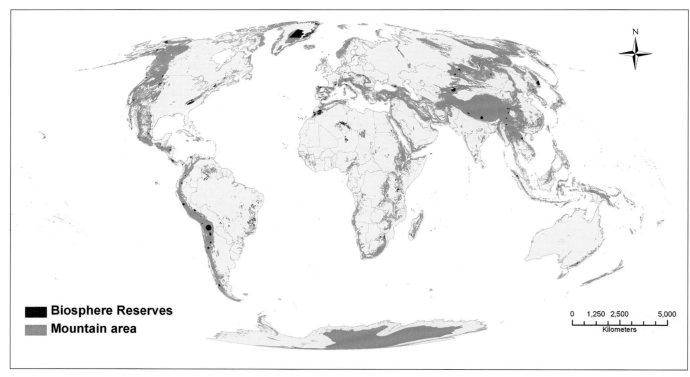

Fig. 2: Distribution of Biosphere Reserves within mountain areas.

of ecosystem services provided by mountain areas would still be hampered by what nowadays represents the major threat to mountains: global change (EEA, 2010a; EEA, 2010b; Kohler & Maselli, 2009). Local or regional adaptation measures to global change in mountains include:

- adaptive management of natural resources;
- connectivity-focused, landscape-scaled land use planning;
- long-term monitoring and research; and
- improved governance, capacity building, and economic support to developing countries with substantial mountain areas under existing schemes for the payment for environmental services (EEA, 2010a; Macchi, 2010).

However, unless urgent, well-targeted measures are undertaken globally to tackle this problem, the impacts of global warming, tourism pressure and land use changes on mountains will only increase, making conservation efforts in these areas largely inefficient (Kohler & Maselli, 2009).

Acknowledgements

We would like to thank our colleagues Simon Blyth and Siobhan Kenney for their helpful comments on an earlier draft of this contribution.

References

EEA (2010a). EEA Report No 6/2010. Europe's Ecological Backbone: Recognising the True Value of our Mountains. Office for Official Publications of the European Union. Luxembourg.

EEA (2010b). EEA Report N° 5/2010. Assessing Biodiversity in Europe: The 2010 Report. Office for Official Publications of the European Union. Luxembourg.

IUCN & UNEP (2010). The World Database on Protected Areas (WDPA): January 2010. UNEP-WCMC. Cambridge, UK.

Kapos, V., Rhind, J., Edwards, M., Prince, M.F., & Ravilious, C. (2000). Developing a map of the world's mountain forests. In Price M.F. & Butt, N. (eds), Forests in Sustainable Mountain Development: A State-of-Knowledge Report for 2000. CABI Publishing. Wallingford, UK.

Kohler, T. & Maselli, D. (eds) (2009). Mountains and Climate Change: From Understanding to Action. Published by Geographica Bernensia with the support of the Swiss Agency for Development and Cooperation (SDC), and an international team of contributors. Bern, Switzerland.

Kollmair, M., Gurung, G., Hurni, K & Maselli, D. (2005). Mountains: Special places to be protected? An analysis of worldwide nature conservation in mountains. International Journal of Biodiversity Science and Management 1, 181–189.

Macchi, M. (2010). Mountains of the World: Ecosystem Services in a Time of Global and Climate Change. ICIMOD. Kathmandu, Nepal.

UNEP-WCMC (2002). Mountain Watch: Environmental Change and Sustainable Development in Mountains. UNEP-WCMC. Cambridge, UK.

UNESCO (1996). Biosphere Reserves: The Seville Strategy and the Statutory Framework of the World Network. UNESCO Man and the Biosphere Programme. Paris, France.

UNESCO (2010). Natural Sciences. Biosphere Reserves at http://portal.unesco.org/science/en/ev.php-URL_ID=4801&URL_DO=DO_TOPIC&URL_SECTION=201.html (visited 13/10/2010).

UNESCO's MAB Programme

The Role of Biosphere Reserves in the Mountain Regions of the World

Ambitious vision and reality

Humans are an integral part of BRs: Cultivation of herbs in Podocarpus BR, Ecuador (© Sigrun Lange).

The Development of UNESCO's MAB Programme, with a Special Focus on Mountain Aspects

by Sigrun Lange

Biosphere Reserves (BRs) are organised under the umbrella of UNESCO's 'Man and the Biosphere' (MAB) Programme. As indicated by the wording, humans play a major role in the development of these sites. Biosphere reserves are not typical nature reserves dedicated mainly to save plant and animal species from human impact. In fact, the conservation of biological diversity is an important objective, but the active role of man as integral part of biosphere reserves, using natural resources sustainably, is an essential factor (at least since the conference in Seville in 1995). BRs are therefore quite often classified as IUCN Category V (protected landscape) (Dudley 2008). However, the concept is not only about preserving cultural and natural landscapes. As model regions for the co-existence of nature and humans, balancing ecological with economic and social needs, biosphere reserves are intended to test and develop future-oriented solutions for today's challenges, such as global warming, loss of biodiversity, or land take-up (for settlements and traffic infrastructure). Much importance is attached to involving local stakeholders in the planning and management of BRs (bottom-up approach). The designation of an area as BR can be perceived as an invitation to society, policy and science to jointly develop wise ways of utilising natural resources, making sure that also following generations will be able to benefit from nature. Summarising, biosphere reserves are to contribute to

- conserving the cultural and biological diversity on earth,
- developing sustainable forms of utilising natural resources (without exhausting them),
- testing new forms of governance by involving local stakeholder groups in discussion and decision-making processes,
- monitoring global change, studying interrelations between man and nature, and facilitating training and environmental education, and
- exchanging experiences and knowledge within the World Network of Biosphere Reserves (WNBR).

Considering this multitude of functions, BRs contribute to the fulfilment of national obligations in the context of international agreements, such as the 'Convention on Biological Diversity', the 'United Nations Framework Convention on Climate Change', and the 'Millennium Development Goals'. Typically, biosphere reserves are divided into three zones, the core, buffer and transition zones (cp. Fig. 1).

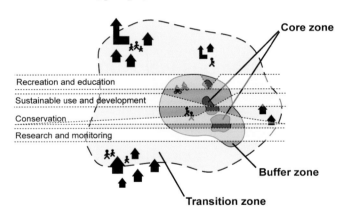

Fig. 1: Zoning in UNESCO biosphere reserves: core, buffer and transition zones (Graph: Sigrun Lange).

- In core zones nature conservation takes precedence. Natural ecosystems must be preserved, human impact is only allowed to a very limited extent. Thus, scientists have the opportunity to study undisturbed dynamic processes in nature. However, in biosphere reserves, the core area often covers only a small part of the total area (unlike national parks, in which 75% of the area is intended to remain in a natural state). The national MAB criteria in Austria and Germany, for example, stipulate five per cent for the core zone (Österreichisches MAB-Nationalkomitee 2005) which is the equivalent of just three per cent (Deutsches MAB-Nationalkomitee 2007) of the total area.
- Buffer zones are to minimise negative external impacts from anthropogenic activities on the core areas. They can also play an important role in promoting connectivity in a larger spatial context as they connect biodiversity components within core areas with those in transition areas (UNESCO 2008).

• People live and make a living in transition zones which are characterised by multiple land use. In innovative pilot projects, sustainable solutions for today's challenges are tested and developed. The boundaries of the transition zone must be clear, whilst cooperation is to extend beyond those boundaries, in order to share best practices, solutions and approaches with the wider region thus fulfilling the role of BRs as learning sites for exemplary sustainable development throughout the region.

The development of the MAB Programme

The new paradigm for the management of protected areas as described by Phillips (2003), can be related to the development of the UNESCO MAB Programme and the purpose of BRs in the past 40 years (cp. Fig. 2). The origin of UNESCO's 'Man and the Biosphere' (MAB) Programme derives from an inter-governmental conference which took place at the UNESCO House in Paris (France) in 1968. The participants in this so-called 'Biosphere Conference' agreed for the first time that biological diversity can be maintained only, if the utilisation and conservation of natural resources go hand in hand rather than in opposition to each other. The Biosphere Conference was the first inter-governmental forum to discuss and promote what is now called 'sustainable development' (UNESCO 1993). One of the twenty recommendations adopted by the participants required UNESCO to set up an international research programme to study the relations between humans and nature, indicating that such a programme should be interdisciplinary in character and take into account the particular problems of developing countries. It was launched in 1971 to develop the basis, within the natural and social sciences, for rational use and conservation of the resources of the biosphere, and for the improvement of the global relationship between people and the environment. In particular, Project 6 of the MAB Programme dealt with the 'Impact of Human Activities on Mountain Ecosystems'. In 1973, an expert group identified the most urgent topics for research in mountain areas as stated below (Schaaf 1999):

• Resource development and human settlements in high tropical mountains, including the tropical Andes, the South Asian mountain complexes and the East African and Ethiopian highlands;
• Tourism, technology and land use in temperate mountains in the middle latitudes, where there are distinct winter and summer seasons;
• Land-use problems in high-latitude mountain and tundra ecosystems, with special reference to grazing, industrial development and recreation.

As a consequence, a large number of case studies was carried out worldwide within the framework of the MAB Programme, in particular in the Andes and the Alps (e.g. impacts of the tourist development on the environment in Obergurgl, Austria). Considerable efforts were made to use standardised methodology for these research activities in order to be able to compare the results on an international level.

	In the past, protected areas were..	Increasingly, protected areas are..
Objectives	• set aside for conservation, • established mainly for spectacular wildlife and scenic protection, • managed mainly for visitors and tourists, • valued as wilderness, • about conservation.	• run with social and economic as well as conservation objectives, • often set up for scientific, economic and cultural purposes, • managed with greater consideration of local people, • valued for the cultural importance of so-called wilderness, • also about restoration and rehabilitation.
Governance	• run by central government.	• run by many partners.
Local people	• planned and managed against people, • managed without regard to local opinions.	• run with, for, and in some cases by local people, • managed to meet the needs of locals.
Wider context	• developed separately • managed as 'islands'	• planned as part of national, regional, and international systems • developed as 'networks' (strictly protected areas, buffered and linked by green corridors)
Perceptions	• viewed primarily as national assets • viewed only as a national concern	• viewed also as community assets • viewed also as international concern
Management techniques	• managed reactively within short time scales • managed in a technocratic way	• managed adaptively with a long-term perspective • managed with political considerations
Finance	• paid for by taxpayer	• paid for from many sources
Management skills	• managed by scientists and natural resource experts • expert-led	• managed by multi-skilled individuals • drawing on local knowledge

Fig. 2: New paradigm for protected areas' management (source: Phillips 2003, p. 20, Tab. 12).

The origin of the World Network of Biosphere Reserves

It was felt that research activities should be concentrated in selected areas. Therefore, the idea was born to set up a worldwide network of so-called 'biosphere reserves', providing the logistic support for repeatable studies. Biosphere reserves were to represent the most important ecosystems and allow for internationally coordinated monitoring of environmental changes. In 1976, the first BRs were nominated (e.g. the mountain areas of Babia Gora BR in Poland or Arasbaran BR in Iran). Most of these areas already had previous conservation designations (e.g. national parks or nature reserves). Since then, the community has grown into a global network of 564 biosphere reserves in 109 countries (UNESCO MAB 2010). Of those almost two thirds are located in mountain regions (cf. Rodríguez-Rodríguez & Bomhard, p. 28).

The 2nd World Congress of Biosphere Reserves in Seville

Even though the concept of biosphere reserves was acknowledged internationally, the quality of the sites differed a lot. A functioning network was a vision rather than reality. The second World Congress on Biosphere Reserves which took place in Seville (Spain) in 1995, marked a milestone in the development of the MAB Programme. The key finding in Seville was that the conservation of biological diversity can no longer be achieved in isolation from the requirements of local people. The active role of humans as an integral part of BRs was emphasised. Local stakeholders are to be involved in planning and management decisions within BRs. UNESCO member states committed themselves voluntarily to meet the requirements expressed in the 'Seville Strategy' and the 'Statutory Framework for Biosphere Reserves'. From that time on, the history of the MAB Programme can be divided in pre- and post-Seville.

In many countries the Seville Strategy stimulated the revision of existing sites, and newly-designated sites corresponded much better to the requirements of 'modern biosphere reserves' than the previous ones. Five years later, during the Seville+5 meeting in Pamplona (Spain) it was observed that local people tended to be more involved in the development processes of BRs.

Fig. 3: UNESCO's MAB Programme recommends a cooperation across borders: Dwarf pine stands in Krkonose/Karkonosze Transboundary BR (© The Krkonoše Mts. NP Administration).

The Pamplona conference was the first to recommend the establishment of transboundary biosphere reserves across national borders. Meanwhile, a total of nine transboundary biosphere reserves (UNESCO 2010) have been designated worldwide, six of them located in mountain regions (e.g. Krkonose/Karkonosze Transboundary Biosphere Reserve, as described by Flousek & Kaspar in this publication; cp. Fig. 3).

The Madrid Action Plan

Since the adoption of the Seville Strategy, new challenges emerged, such as global warming in particular, the accelerated loss of biological and cultural diversity, and the increasing urbanisation. In 2008, during the 3rd World Congress of Biosphere Reserves in Madrid (Spain), the Madrid Action Plan 2008–2013 was adopted, inviting the members of the World Network of Biosphere Reserves (WNBR) to deal with mitigation and adaptation measures related to global warming, and to establish themselves as exemplary learning sites for sustainable development. The significance of buffer and transition zones was highlighted; they were to be increased by '*linking up relatively small protected core areas with significantly larger buffer zones and transition areas*' (UNESCO 2008, p.18).

Finally, it can be pointed out that biosphere reserves have evolved from mere research sites in the 1970s and 1980s into important key players in the 21st century who are achieving the challenging task of putting into practice the abstract concept of sustainable development both at local and regional level (cf. Fig. 4). They are the only protected area category which makes it possible to link a wide range of different living spaces, ranging from untouched, strictly protected natural landscapes in core zones to strongly impacted, even urban areas in transition zones where people make a living, with the objective to do it differently, more innovatively, more sustainably, more focused on a sustainable future. As reflected by the vision expressed in the Madrid Action Plan (UNESCO 2008, p.8), the '*World Network of Biosphere Reserves (WNBR) consists of a dynamic and interactive network of sites of excellence. It fosters harmonious integration of people and nature for sustainable development through participatory dialogue, knowledge sharing, poverty reduction and improvement to human well-being, respect for cultural values and society's ability to cope with change, thus contributing to the UN Millennium Development Goals. Accordingly, the WNBR is one of the main international tools which facilitate the design and implementation of sustainable development approaches in a wide array of contexts*'.

So much for the vision – the challenging task of implementing UNESCO's MAB concept in an ideal manner may be a little over-ambitious. The reality shows a different picture. In this publication, Borowski and Munteanu (p. 39) conclude that '*at the moment, many [biosphere reserves] appear to be not much more than paper reserves*'. Other critical articles were submitted for this publication by Vološčuk (p. 45); Schmidt (p. 75); Saxena, Maikhuri and Rao (p. 79); and Boussaid (p. 83).

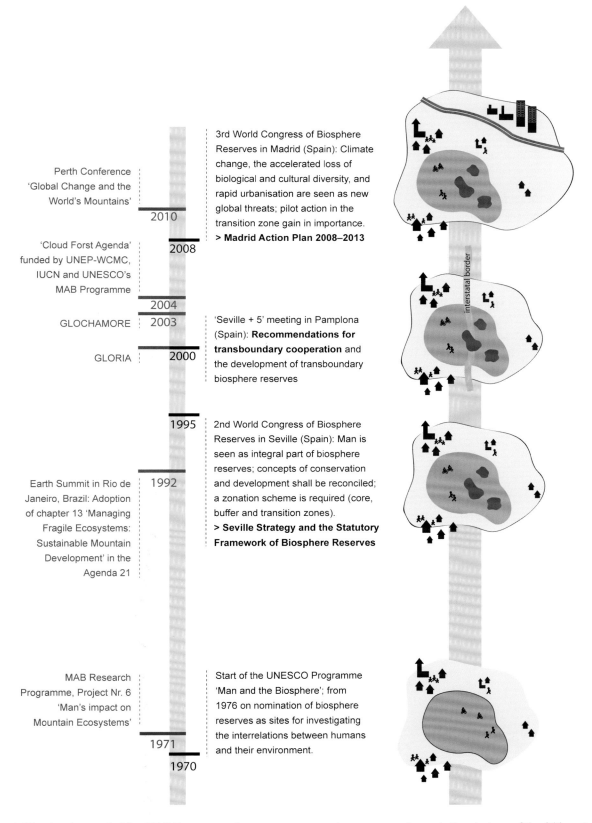

Perth Conference
'Global Change and the
World's Mountains'

2010

3rd World Congress of Biosphere
Reserves in Madrid (Spain): Climate
change, the accelerated loss of
biological and cultural diversity, and
rapid urbanisation are seen as new
global threats; pilot action in the
transition zone gain in importance.
> Madrid Action Plan 2008–2013

2008

'Cloud Forest Agenda'
funded by UNEP-WCMC,
IUCN and UNESCO's
MAB Programme

2004

GLOCHAMORE

2003

'Seville + 5' meeting in Pamplona
(Spain): **Recommendations for
transboundary cooperation** and
the development of transboundary
biosphere reserves

GLORIA

2000

1995

2nd World Congress of Biosphere
Reserves in Seville (Spain): Man is
seen as integral part of biosphere
reserves; concepts of conservation
and development shall be reconciled;
a zonation scheme is required (core,
buffer and transition zones).
**> Seville Strategy and the Statutory
Framework of Biosphere Reserves**

Earth Summit in Rio de
Janeiro, Brazil: Adoption
of chapter 13 'Managing
Fragile Ecosystems:
Sustainable Mountain
Development' in the
Agenda 21

1992

MAB Research
Programme, Project Nr. 6
'Man's impact on
Mountain Ecosystems'

Start of the UNESCO Programme
'Man and the Biosphere'; from
1976 on nomination of biosphere
reserves as sites for investigating
the interrelations between humans
and their environment.

1971

1970

interstatal border

Fig. 4: The development of the MAB Programme from a mere research programme towards the strategy of the 21st century for sustainable development (Graph: Sigrun Lange). [Black writing: Milestones of the MAB Programme; green writing: activities focussed on mountain issues]

Regional networks

The WNBR is furthermore structured into regional and thematic networks to facilitate cooperation amongst like-minded parks. EuroMAB, for example, is the largest and oldest of the regional networks. It comprises 262 BRs in 52 countries, including Canada and the USA. Since 1986, EuroMAB meetings take place approximately every two years. The last one was held in Tatry Biosphere Reserve in Slovakia in October 2009. Another network, IberoMAB, aims at strengthening biosphere reserves in Latin American and Caribbean countries, Spain and Portugal, notably by consolidating their MAB National Committees, and promoting the creation of new biosphere reserves. The 14th IberoMAB meeting will be held jointly with the conference on 'Biosphere Reserves: All Hands, All Voices' in Puerto Morelos, Mexico, in November 2010.

Thematic focus on mountain issues

UNESCO's MAB Programme is engaged in fragile ecosystems, such as mountain regions, threatened by climate change, species extinction, natural hazards (erosion and floods), and changing land use which modifies the socio-economic conditions and livelihoods of people. In particular, UNESCO-MAB assesses the impacts of global change on fragile mountain ecosystems by using mountain biosphere reserves as study and monitoring sites (cf. Lange, p. 50, p. 61) and organising conferences on mountain topics ('Global Change and the World's Mountains' in Perth, Scotland, October 2010). In addition, UNESCO has established two Chairs in Sustainable Mountain Development, one at the International University of Kyrgyzstan held by Asylbek Aidaraliev, and another one at the Centre for Mountain Studies at UHI-Perth College, United Kingdom, held by Martin Price. It is to be hoped that the activities carried out in mountain biosphere reserves by various key players will contribute to the overall goal of conserving these fragile and valuable ecosystems in the long-term.

References

Bubb, P., May, I., Miles, L. & Sayer, J. (2004). Cloud Forest Agenda. UNEP-WCMC, Cambridge, UK.

Deutsches MAB-Nationalkomitee (Hrsg.) (2007). Kriterien für die Anerkennung und Überprüfung von Biosphärenreservaten der UNESCO in Deutschland.

Dudley, N. (ed.) (2008). Guidelines for Applying Protected Area Management Categories. IUCN, Gland, Switzerland.

Lange, S. (2005). Leben in Vielfalt. UNESCO-Biosphärenreservate als Modellregionen für ein Miteinander von Mensch und Natur. Der österreichische Beitrag zum UNESCO-Programm „Der Mensch und die Biosphäre". Verlag der Österreichischen Akademie der Wissenschaften.

Österreichisches MAB-Nationalkomitee (Hrsg.) (2005). Nationale Kriterien für Biosphärenparks in Österreich.

Phillips, A. (2003). Turning Ideas on Their Head. The New Paradigm for Protected Areas. In: The George Wright FORUM.

Schaaf, T. (1999). UNESCO's Man and the Biosphere Programme in mountain areas. In: Unasylva 196. Download at: http://www.fao.org/docrep/x0963e/x0963e08.htm#TopOfPage (accessed on 26 October 2010)

UNESCO, MAB (ed.) (2010). Official web site of UNESCO's MAB Programme at: http://www.unesco.org/mab (accessed on 20 November 2010)

UNESCO, MAB (ed.) (2008). Madrid Action Plan for Biosphere Reserves (2008 – 2013).

UNESCO, MAB (ed.) (2003). Five Transboundary Biosphere Reserves in Europe. Biosphere Reserves Technical Notes. UNESCO, Paris.

UNESCO, MAB (ed.) (2000a). Seville+5 Recommendations: Checklist for Action.

UNESCO, MAB (ed.) (2000b). Seville+5 Recommendations for the Establishment and Functioning of Transboundary Biosphere Reserves ('Pamplona Recommendations').

UNESCO, MAB (ed.) (1995). 'The Seville Strategy for Biosphere Reserves' and 'The Statutory Framework of the World Network of Biosphere Reserves'.

UNESCO, MAB (1993). The Biosphere Conference, 25 years later.

Triglav National Park, Slovenia, is also nominated as Biosphere Reserve (© Sigrun Lange).

Biosphere Reserves in European Mountains: an Exploratory Survey

by Diana Borowski & Catalina Munteanu

Introduction

Mountain areas are valuable centres for biodiversity, cultural diversity and traditional ecological knowledge. They include a wide range of ecosystems which provide critical goods and services to both mountain and lowland people. Mountains are Europe's water towers and key locations for tourism and recreation (EEA 2010). In a globalising world, mountain areas and their people face increasing challenges and, in this context, are ideal locations for the implementation of the Biosphere Reserve (BR) concept. Nevertheless, owing to the diversity, difference in size, history of designation, income sources and potentials, the effectiveness of mountain BRs varies strongly from one region to another. Although some positive impacts of BR designation in Europe's mountains can be observed, the implementation of the Seville Strategy, the Statutory Framework of the World Network of Biosphere Reserves (below, 'the Statutory Framework') (UNESCO 2005) and the Madrid Action Plan (UNESCO 2008) is still not complete in most of these BRs.

Methodology

Owing to major differences between Europe's mountain BRs, generalisations can be made only to a very limited extent. The main information source for the map of these BRs was the 2010 WDPA Annual Release[1], and the delineation of European mountains is based on the ETC-LUSI/EEA (November 2008).[2] The other aspects addressed below reflect the information available in a selection of publications and other documentation on mountain BRs, complemented by phone interviews with protected-area managers. The overall picture is not exhaustive, but provides some insights and examples. What is the spatial distribution of BRs in Europe's mountains? Do they have an impact on sustainable development and on biodiversity conservation, or are they just a concept?

The assessment of the effectiveness of designation as a BR proved to be challenging. We established a set of ten 'indicators of effectiveness' in order to outline 'good-practice' and 'less effective' examples of BR implementation in Europe's mountains. The choice of examples is based on a geographical balance of European mountain ranges and on information availability. The indicators reflect in a generalised manner the criteria of the Seville Strategy and the Statutory Framework and aim to depict some of the main prerequisites for BRs to be effective: the inclusion of the concept of BRs in national legislation; the existence of a dedicated management plan as well as adequate funding and a separate management structure (i.e. an authority that is *only* responsible for the BR); the involvement of the local population; research activities undertaken within the BR; and the development of an 'image' (e.g. a logo and/or a website and/or 'brands' for products produced within the reserve). Based on these indicators, we could identify examples of effective and less effective BRs which can then serve as models for future developments and designations.

Location and distribution

At present, 84 of Europe's 158 BRs[3] are located within mountain ranges[4]. The Alps – Europe's highest and most famous mountain range – have a comparatively low number of BRs: two in the Swiss, three in the Austrian, one in the Slovenian, one in the German, and none in the Italian Alps. Not a single BR is located in the Scandinavian mountains (Norway, Sweden, or Finland). The only Swedish mountain BR (Lake Torne Area, designated in 1986) was withdrawn from the UNESCO list in June 2010 since it did not fulfil the criteria of the Statutory Framework. Both Spain and Bulgaria host a remarkably high number of mountain BRs. All of the eleven BRs in the Bulgarian mountains were designated in 1977, and all are *'currently under revision'* according to UNESCO's website.[5] In addition, no

---

[1] http://www.wdpa.org, accessed on 4.10.2010
[2] Delineation used for the EEA report 2010

[3] This excludes the Russian Federation and Israel.
[4] This excludes the biosphere reserve North-East Greenland.
[5] http://www.unesco.org/mab; however it must be noted that the 'MAB Biosphere Reserves Directory' on the UNESCO website has in many respects not been updated since 2002.

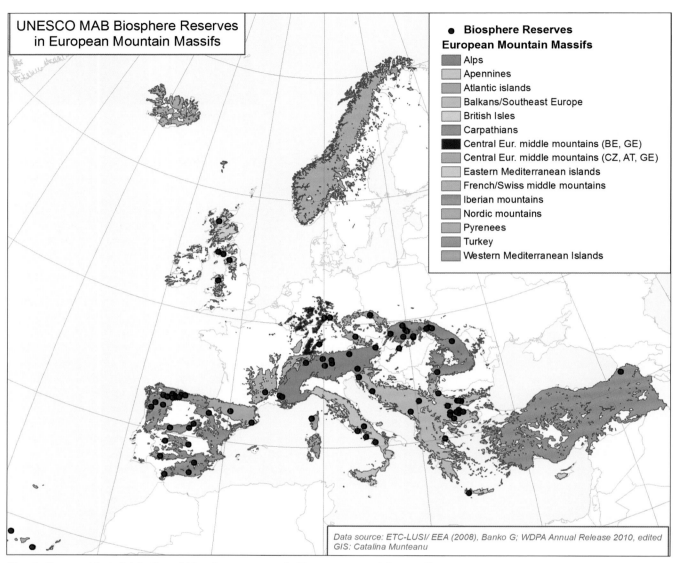

UNESCO MAB Biosphere Reserves
in European Mountain Massifs

● **Biosphere Reserves**
European Mountain Massifs

- Alps
- Apennines
- Atlantic islands
- Balkans/Southeast Europe
- British Isles
- Carpathians
- Central Eur. middle mountains (BE, GE)
- Central Eur. middle mountains (CZ, AT, GE)
- Eastern Mediterranean islands
- French/Swiss middle mountains
- Iberian mountains
- Nordic mountains
- Pyrenees
- Turkey
- Western Mediterranean Islands

Data source: ETC-LUSI/ EEA (2008), Banko G; WDPA Annual Release 2010, edited
GIS: Catalina Munteanu

Map 1: Geographical distribution of biosphere reserves in European mountain massifs.

periodic review report has been submitted for any of them (Price, Jung & Bouamrane 2010). By contrast, of the 27 BRs in Spanish mountain ranges, 18 were designated after 1995 i.e. after the adoption of the Seville Strategy. It is more likely that they fulfil the criteria for BRs as set out in Seville. However, detailed information about most of the issues examined in this paper is also extremely sparse[6].

The only transboundary BRs in Europe's mountains could until recently be found in Poland, the Czech Republic, Slovakia and the Ukraine: the Tatry, Krkokonose/Karkonosze and the East Carpathians BRs. In 2009, the Geres/Xures BR was designated on the Spanish/Portuguese border.

The British, Greek, Turkish and Romanian mountain ranges each contain one to five BRs, as do the German uplands, the Apennines, and the Massif Central. All BRs in Europe's mountains coincide with at least one other type of protected area: a very large number incorporate a national park, but other designations incorporate (regional) 'Naturparks' (wildlife parks), nature reserves, Ramsar wetlands and, of course, Natura 2000 sites.

Implementation of the Seville Strategy and contribution to biodiversity conservation and sustainable development
The BRs in Europe's mountain ranges differ quite significantly in their implementation of the 'size' and 'zonation' criteria of the Statutory Framework. To generalise: the later a BR was designated, the more likely it is that these criteria are being implemented. Austria and Switzerland are a case in point. In Austria, the Gurgler Kamm and Gossenköllesee BRs, both designated in 1977, do not fulfil the requirements, as they are too small and do not feature the required zonation (Lange 2005).

---

[6] The most comprehensive source is Anon (2006): La Red de Reservas de la Biosfera Españolas. Lunwerg Editores, Barcelona/Madrid.

By contrast, the Grosses Walsertal BR, established in 2000, is often named as an example of 'best practice'. In Switzerland, the 'active' biosphere reserve Entlebuch, designated in 2001, felt the need to underline their difference from the 'passive' biosphere reserve Parc Suisse (established in 1979, focus on wilderness conservation) on its website.[7] However, this might change in the future, as the later one was extended and renamed in 2010, now including approx. 1,600 residents of the Val Müstair who in a public referendum voted with a vast majority (89%) in favour of being included in the BR.

A number of Europe's mountain BRs would not receive this designation under the criteria in place since 1995. This corresponds to the findings in the Madrid Action Plan, adopted by UNESCO in 2008: *'98% of the places nominated as biosphere reserves since 1995 have adopted the three-zone scheme. For those biosphere reserves included in the WNBR prior to 1995, this percentage was 23% for those between 1976 and 1984, and 65% for others designated during 1985–1995'.* When analysing the impact of Europe's mountain BRs on biodiversity conservation and sustainable development, the overall verdict is disappointing (Table 1). For many of the older

BRs (established in the 1970s), information is hard to obtain. Since many do not even implement the most basic criteria of the Statutory Framework (adequate size and three zones), it seems reasonable to assume that their impact is low. However, even the more recently designated BRs are not necessarily successful in reaching their goals.

'Less effective' biosphere reserves

In Slovenia, the Julian Alps BR has been more or less inactive since its designation in 2003. Many local people are not aware that they live within a BR. There is no management structure exclusively for the BR; the administration of Triglav National Park (which makes up the core zone) is also responsible for looking after the BR. Therefore the focus of development has been the core zone, but no progress was made in the transition zone. So far, there is neither a logo nor a website dedicated to this BR. One reason for the lack of involvement of the local population is that the BR was mainly a 'top-down' project: It was initiated by UNESCO Slovenia, the Ministry for Spatial Planning, and the management of Triglav National Park, but local people were not asked for their opinion (Del Negro 2009).

---

[7] Web site of the Entlebuch BR: http://www.biosphaere.ch

Effectiveness Indicators	Year of designation	Category in national legislation	Management Plan	Funding *	Zonation	Adequate size	Authority: BR Manager	Involvement in research/ education **	Involvement of local communities	Local branding/ image development
Camili BR (Turkey)	2005	✘	?	?	✘	?	?	✓	✓	?
Julian Alps BR (Slovenia)	2003	✘	✘	✘	✓	✓	✘	✓	✘	✘
Dehesas de Sierra Morena BR (Spain)	2002	?	?	?	✓	✓	?	✓	?	?
Entlebuch BR (Switzerland)	2001	✘	?	✓	✓	✓	✓	✓	✓	✓
Golja Studenica BR (Serbia)	2001	✘	✘	?	✓	?	✓	✓	✘	✓
Großes Walsertal BR (Austria)	2000	✓ (b)	✓	✓	✓	✓	✓	✓	✓	✓
East Carpathians BR (Ukraine)	1998	✘	✘	?	✓	✓	?	?	?	?
Rhön BR (Germany)	1991	✓	✓	✓	✓	✓	✓	✓	✓	✓
Sumava BR (Czech Republic)	1990	✘	✘	?	✓	✓	✘	✓	✘	✓
Samaria Gorge BR (Greece) (a)	1981	✘	✘	?	✓	?	?	✓	?	✘
Pietrosul Mare BR (Romania)	1979	✓	✓(c)	✘	✘	✘	✘	✓	✘	✘
Boatin, Tsarichina, Steneto, Djendema BRs (Bulgaria)	1977	?	✘	?	✘	✘	✘	?	?	?
Babia Gora BR (Poland)	1976	✘	✘	?	✓	✓	✘	?	✘	✘

Tab. 1: Effectiveness indicators for biosphere reserves.
* independent from national park; ** includes activities in national parks
(a) multi-designation of the site makes the distinction of BR-specific indicators difficult
(b) BR is a category in the law of the federal-state Vorarlberg
(c) there are no communities in the BR, the management plan exists as part of the National Park Plan

Similar problems are faced by Babia Góra BR in Poland: lack of a coordinating body (the BR is managed by the Babia Góra National Park administration), weak collaboration with municipal authorities, inadequate funding, and a shortage of personnel (Schliep & Stoll-Kleemann 2010, p. 924). This, however, is not due to legislative reasons. On the contrary, the Polish government has adopted a supportive national nature conservation policy – but this policy only pertains to national parks and other protected areas; BRs are not recognised as a category in the context of this policy (Schliep & Stoll-Kleemann 2010, p. 923).

The East Carpathians BR, designated in 1998, is the first tri-national BR in the world. Challenges of international cooperation have made its operation even more complex: To date, the governments of Poland, Slovakia and the Ukraine have not signed an official letter of agreement on the BR, nor does a joint management plan exist. Inconsistent BR zoning schemes in each of the three countries make coordination difficult, and research activities are set up exclusively at a national level (Bihun et al. 2008, p.1/2).

The existing BRs in Bulgaria are from the 'pre-Seville' generation, focusing mainly on conservation and research, lacking zonation and featuring very small areas of integral protection. They are included within other categories of nationally designated sites and lack individual management structures. For example, within the Central Balkan National Park, four of the most strictly protected reserves (Boatin, Tsarichina, Steneto and Djendema) are designated as BRs: they are managed under a strict conservation regime and closed for any activities except for limited research and therefore do not precisely meet the criteria of the Statutory Framework (BSBCP). The way forward for these small protected areas would be to include them in a larger, adequately-sized and zoned reserve; alternatively, withdrawal of the UNESCO status could be envisaged. Within its management plan[8], approved by Ministers in 2001, the Central Balkan National Park (including the current BRs) has been proposed for designation by MAB but there is no recent information on progress in this matter.

Romania hosts two 'pre-Seville' mountain BRs: Muntii Rodnei (overlapping partly with the Retezat National Park) and Pietrosul Mare (its boundaries matching those of the Rodna Mountains National Park). The latter was designated in 1979 and, until recently, did not comply with most principles of the Seville Strategy. In 2002, the reserve was reviewed and extended (to an area of 44,000 ha) but it is still not zoned. Additionally, the transition period in Romania since 1990 (with the collapse of the mining industry, migration to Western Europe) still poses a great economic and social challenge to the sustainable development of the region. Understanding of the concept and local awareness of the existence of the BR is still very low. The Romanian legislation offers a fairly adequate framework for BRs, which are specified as a category in the Act No. 462/2001 on Protected Areas. The Government's Decision No. 230/2003 details the boundaries and management structure

of Pietrosul Mare BR as part of the Rodna Mountains National Park. Currently, the reserve is managed within the park, having a designated chapter in the management plan. The park's representatives have recently started a consultation process with national authorities and local communities for the designation of a management authority and the extension and zonation of the BR to an area of approx. 134,000 hectares, to encompass adjacent settlements (Iusan 2006). Problems faced in this process include the continued dominance of top-down approaches, and both, lack of funding methodological know-how and communication.

In Greece, Samaria National Park (White Mountains) has many designations: Biosphere Reserve, National Park, Council of Europe Diploma Site, Natura 2000 site and wildlife reserve. The requirements of most designations were traditionally central management by a legally designated body. The multiple designations, however, do not necessarily bring immediate benefits to the area. The hierarchy and the decision-making process, and the lack of specified management bodies, action plans, authority and resources for the fulfilment of each designation, pose serious problems for the Samaria BR. After 1999, the implementation of the Seville Strategy was slowed down by a transitional period in the legislative process. However, after 2004, these legal changes influenced the management of the multi-designated area towards conservation, community development and logistic support in a fairly positive way (Kargiolaki 2005).

'Good practice' biosphere reserves

In contrast with the examples described above, several BRs are repeatedly named as examples of 'good practice' throughout the literature, including Entlebuch (CH), Rhön (DE) and Großes Walsertal (AT). An example of particularly strong involvement of the local population is the Swiss Entlebuch BR. It was the first BR anywhere created by means of a referendum: in 2000, the eight communities concerned approved the proposed reserve with an overwhelming majority of 94 per cent. Nowadays, governance is realised through a system of regional management with public participation. Representatives of the individual towns and various organisations are elected to a steering committee by an assembly of delegates (Hambrey Consulting 2007). In Entlebuch, the conditions for achieving sustainable regional development via a BR exist: the stakeholders involved participated in the formulation of objectives, the biosphere management was granted the requisite authority, the necessary funds were made available, and institutional adaptations were made (Hammer 2007). Positive results were achieved on many levels: According to the website, biodiversity in the reserve has stabilised and the population of endangered species increased. There is reported to be an increased identification with natural and cultural values, leading to reinforced regional self-confidence. Numbers of visitors, meals and over-night stays have increased since the reserve was designated, and it is reasonable to assume that these changes can at least partly be credited to the BR. The creation of the 'Biosphäre' has also coincided with an increase in 'eco-tourism' (Hammer 2007).

---

[8] http://www.centralbalkennationalpark.org (accessed on 10-10-2010)

The Rhön BR was designated immediately after the German unification and comprises administration units from three provinces (Länder), which coordinate research, tourism development, land use planning and projects. Local people and local businesses are represented in different bodies within the management structure. A couple of initiatives were launched for the marketing of regional products, including the Rhön sheep and Rhön apples. Even though the effect (for example in terms of tourist numbers) is hard to quantify owing to lack of statistics (Hambrey Consulting 2007), members of the management team claim that the activities in the BR have had a positive impact on the local economy, leading for example to an increase in sales of regional products and ensuring conservation of genetic resources. In addition, the activities in the BR created a strong regional identity and enhanced cooperation between local entrepreneurs (Pokorny). The Großes Walsertal Biosphere Park was designated in 2000 and a manager was appointed. It has initiated a number of projects with the aim of the sustainable development of the regional economy: promotion of organic farming, marketing of local agricultural products (brands such as 'Walserstolz', 'Bergtee', 'die köstliche Kiste'), the label 'Partner Company of the Biosphere Beserve' for restaurants and lodges, and a label for timber from the biosphere park. A study of the impact of these initiatives (Coy & Weixl-baumer 2005) draws an encouraging conclusion: It found, for example, that the label 'Partner Company of the Biosphere Reserve' was viewed favourably by the participating companies, but that awareness among tourists remained insufficient. The number of tourists in the area had not significantly increased since the designation of the BR (only 1/4 of existing lodges had noted increases) but at least tourist numbers had not declined either. Most visitors knew about the BR, but only very few named the designation as 'crucial' in the choice of their holiday resort.

Discussion

When analysing the 'factors of success' of these areas, one clear conclusion is that the involvement of the local population is of major importance, as is the establishment of a separate management structure specifically dedicated to the BR. In addition, the development of brands for local products seems to be a good idea to raise awareness for the products, to create an image for the region and ultimately to attract tourists. However, these brands must be advertised as widely as possible, otherwise awareness of the brand will remain low. Nevertheless, even this generally positive assessment of the examples cited above is subject to reservations. For example, the Grosses Walsertal Biosphere Park lacks sufficient funding, and the park manager has to accomplish, on her own, the same workload shouldered by over 30 employees in the neighbouring Kalkalpen National Park (Lange 2005).

While national parks (or similar categories of protected areas) are usually recognised in national legislation and provided with at least a basic amount of funding, BRs are often not recognised in legislation. Despite designation by UNESCO, they receive no funding from the international community. They are financed principally by the national or sub-national governments within the countries in which they are situated; they can try to attract external funding (from various EU programmes, from NGOs, local businesses, etc); or they receive no funding at all. Consequently, many countries do not even put into place a separate management structure but simply add the management of the BR to the list of responsibilities to be borne by national park managers, given that most BRs overlap with a national park.

For many of the reserves, the strict nature conservation role (generated in the first, 'pre-Seville' phase of designation) is still prominent (Price 2002). The core area of BRs either overlaps or overlays an existing designated protected area, and new land or new functions hardly add value to the designation. Thus, BRs in European mountains would be able to contribute to sustainable development: in contrast with many other categories of protected areas which focus exclusively on biodiversity protection, the holistic approach of BRs extends the idea of sustainability to society, culture and economy, accommodating needs of both man and the environment. But at the moment, many appear to be not much more than 'paper reserves'. Off the record, some protected area managers from the examples described above admit that the MAB label is just a 'cosmetic add-on' without content in many national parks (Schliep & Stoll-Kleemann 2010, p. 925). This may be the symptom of a fundamental misunderstanding of the MAB logo as a 'label', as identified by Nolte: *'Outsiders to the program (and sometimes even insiders) often fail to see that – in contrast to a World Heritage designation that can be understood as a recognition for something outstanding that already exists – being a BR only refers to an intention, a commitment to networking, participation, exchange and sustainable development, and as such is not a guarantee for actual positive changes in management practice'* (Nolte 2006, p.5).

To address these challenges, UNESCO might raise the profile of BRs by more strict enforcement of the criteria in the Statutory Framework, which states that a BR can be removed from the network if it does not satisfy the criteria for designation. Since 1996, a periodic review process has been in place for BRs: The Statutory Framework requires a report on the status of BRs every ten years. So far, 229 periodic review reports have been submitted. But 130 reports have not been submitted for BRs designated before 2000; and one fifth of the countries with BRs have never submitted a periodic review report. Ten BRs have so far been withdrawn from the WNBR by states voluntarily, but none have been removed at UNESCO's initiative (Price, Park and Bouamrane 2010, p. 552/553). A more stringent handling of the cases of BRs that do not fulfil the criteria might give more credibility to the whole concept. Retaining mismanaged BRs in the network damages the reputation of the MAB Programme, and by extension, the reputation of UNESCO itself (Nolte 2006, p.5). To complement this, UNESCO might increase awareness of examples of 'good practice' and formally recognise those responsible for BRs for such endeavours. This could greatly add to their value both as a concept and for ensuring that biodiversity conservation is directly linked to sustainable development in practice, not only in Europe's mountain BRs, but throughout the World Network of BR.

References:

Adem, C. et al (2007). Governance and Ecosystems Management for the Conservation of Biodiversity. GEM-CON-BIO Case Study Report Camili Biosphere Reserve Turkey, FP6 Project no: 028827. Available at: http://www.gemconbio.eu/downloads/GEMCONBIO%20case%20study%20report%2010.2%20-%20Camili%20Biosphere%20Reserve%20Turkey.pdf

Bihun, Y.M., Keeton, W.S., Stankiewicz, O. & Ceroni, M. (2008). Transboundary Protected Areas Cooperation in the East Carpathian and Carpathian Biosphere Reserves – Final Report. WWF Österreich.

BSBCP - Bulgarian-Swiss Biodiversity Conservation Programme (year unknown). Management plan for the highland treeless zone of the Central Balkan National Park, Volume 1. Available at: http://www.bbf.biodiversity.bg/files/doc/cb_en_mplan_vpblzona.pdf

Coy, M. & Weixlbaumer, N. (2005). Zukünftige Entwicklungsstrategien für den Biosphärenpark Großes Walsertal – Eine regionalwirtschaftliche und perzeptionsgeographische Analyse. Projekt zum Aufruf 'Forschung an der Schnittstelle zwischen Natur- und Sozialwissenschaften' der ÖAW zum Man and Biosphere Programme.

Del Negro, M. (2009). Schutzgebiete in Slowenien mit besonderer Berücksichtigung des Alpenraums – Bestandsaufnahme und Problematik. IGF-Forschungsbericht, Band 3, Verlag der Österreichischen Akademie der Wissenschaften.

EEA (2010). Europe's ecological backbone: recognising the true value of our mountains. EEA report No 6/2010. European Environmental Agency, Copenhagen.

Hambrey Consulting (2007). Social, economic and environmental benefits of World Heritage Sites, Biosphere Reserves, and Geoparks. Scottish Natural Heritage Commissioned Report No.248. Available at: http://www.snh.org.uk/pdfs/publications/commissioned_reports/Report%20No248.pdf

Hammer, T. (2007). Biosphere Reserves: An Instrument for Sustainable Regional Development? The case of Entlebuch, Switzerland. In: Mose (ed.): Protected Areas and Regional Development in Europe. Ashgate Pub Co, Aldershot.

Instituto de Ecología (2008). Dehesas de Sierra Morena. Ficha de Reserva. Contribution for the website of the institute. Available at: http://proyectos.inecol.edu.mx/dms/Documents/Fichas_de_Reservas/Espana/Espana%201/RB_DehesasDeSierraMorena_ES.pdf

Iusan, C. (2006). Evaluarea situatiei actuale a managementului Rezervatiei Biosferei Pietrosul Mare, Acta Musei Maramorosiensis IV, pp. 155–161.

Kargiolaki, H. (2005). White Mountains (Samaria Gorge), focus on the management aspects according to the various designations (Biosphere Reserve, National Park, Council of Europe Diploma Site, Natura 2000 site and the wildlife refuge), In: EUROMAB AUSTRIA 2005 Meeting of the Euromab Biosphere Reserve Coordinators and Managers, Proceedings.

Kargiolaki, H. (2005). Biosphere Reserve of Samaria: difficulties faced in introducing a new management body. In: EUROMAB AUSTRIA 2005 Meeting of the Euromab Biosphere Reserve Coordinators and Managers, Proceedings.

Lange, S. (2005). Leben in Vielfalt – UNESCO Biosphärenreservate als Modellregionen für ein Miteinander von Mensch und Natur. Verlag der Österreichischen Akademie der Wissenschaften, Wien.

Vázquez, M., Villa Díaz, F. & A. (2002). Dehesas de Sierra Morena: La octava Reserva de la Biosfera de Andalucía. In : Revista Medio Ambiente No. 41. Available at : http://www.juntadeandalucia.es/medioambiente/contenidoExterno/Pub_revistama/revista_ma41/ma41_14.html

Nolte, Ch. (2006). The Biosphere Tour - Report to the MAB Secretariat. UNESCO field offices and Biosphere Reserves, Berlin, 21.08.2006. Available at: http://www.biosphere-tour.org/files/biosphere_tour_2005_2006_report.pdf

Pokorny, D. (year unknown). Biosphere Reserve and Local Economies – Case Study: Rhön Biosphere Reserve, Germany. Available at: http://www.ecodyfi.org.uk/biosphereproject/downloads/doris.doc

Popesku, I. (2002). The tourism potentials and impacts in protected mountain areas 'Golija-Studenica' Biosphere Reserve, Serbia. Case Study. In: International Workshop for CEE Countries. 'Tourism in Mountain Areas and the Convention on Biological Diversity'. 1-5 October 2002, Sucha Beskidzka, Babia Gora National Park, Poland.

Price, M., Jung J.P. & Bouamrane, M. (2010). Reporting progress on internationally designated sites: The periodic review of biosphere reserves. In: Environmental science & policy 13; 549–557.

Price, M. (2002). The Periodic Review of Biosphere Reserves: A mechanism to foster sites of excellence for conservation and sustainable development. *Environmental Science and Policy* 5(1): 13–19.

Schliep, R. & Stoll-Kleemann, S. (2010). Assessing governance of biosphere reserves in Central Europe. In: Land Use Policy 27; 917–927.

UNESCO (2008). Madrid Action Plan for Biosphere Reserves (2008–2013). UNESCO, Paris.

UNESCO (2005). Biosphere Reserves: The Seville Strategy and the Statutory Framework of the World Network. UNESCO, Paris.

UNESCO (1995). The Statutory Framework of the World Network of Biosphere Reserves. UNESCO, Paris.

Impressions from Tatry National Park and Biosphere Reserve (© Juraj Ksiažek).

The MAB Programme – Vision and Reality: Case Study of the Transboundary Tatry Biosphere Reserve

by Ivan Vološčuk

Objectives of UNESCO´s MAB Programme

Biosphere reserves are required to meet high standards: they are considered to be sites of excellence with the aim to reconcile the conservation of biodiversity with economic development. Naturally, this is not an easy task. Many areas struggle to pursue conservation and development in equal measure. At the end of the day, economy often overrules ecology – no matter how many prestigious international labels a conservation area has collected. The case study of Tatry National Park and Biosphere Reserve illustrates that even recognised international labels such as the ones designated by IUCN and UNESCO, are no guarantee that ecologically valuable areas are prevented from degradation by ongoing human development.

Tatry Mountain's biological and cultural significance

The Tatry Mountains (West Tatry, High Tatry and Belianske Tatry Mountains) are situated in the northern part of the Slovak Republic. They are the highest mountains in the Carpathian range which extends over 1,800 kilometres from Slovakia into Romania, via Poland, Ukraine and Hungary. They are an outstanding crossroad on the migration routes of alpine and arctic biota – an 'island' where rare mountain ecosystems with unique plant and animal species are preserved. A total of 1,300 species of vascular plants occur in the Tatry Mountains. These include several species of Nordic origin, most of which are relict species from the last ice age at the southern edge of their distribution in the Tatry, and several Tatran and Carpathian endemics. More than 150 species are listed for protection. Among the most endangered ones are edelweiss *(Leontopodium alpinum)*, pasqueflower *(Pulsatilla alba* and *P. vernalis)*, and moor-king *(Pedicularis sceptrum-carolinum)*. Eight reptile, three amphibian, 115 bird and 42 mammal species are found there. Carnivores include brown bear, northern lynx, wolf, common wild cat, pine marten, common badger, and other more common species. However, apart from their biological significance, the Tatry Mountains are also a symbol of national identity and independence of the indigenous Slavonic people who settled in the West Carpathian basins many centuries ago. With currently about half a million visitors, it is Slovakia's most popular tourist destination.

Conservation efforts

With the objective of preserving this outstanding highland area, the area of Tatry Mountains was declared the first national park in Slovakia (Tatranský národný park, TANAP) in 1949. Five years later, in 1954, Tatrzański Park Narodowy (TPN) was established on the Polish side. In the 1980s, the national park councils on both sides of the border started to develop a concept for a future joint protected area covering fragments of both countries. UNESCO´s 'Man and the Biosphere' Programme offered the only existing internationally recognised model for transboundary conservation units. Consequently, the Tatry BR was approved in 1993. It covers two national parks on either side of the political boundary between Poland and Slovakia. The Slovak part accounts for three quarters of the total area. Although being classified as Category II of the IUCN Protected Areas Category System, the two sites have very different administrative structures.

In parallel with the development of Tatry Biosphere Reserve, the 'Management Programme for TANAP until 2000' was approved by the government of the Slovak Republic in 1991, in order to achieve the goal of preserving the irreplaceable natural resources of the Tatry. This Programme was based conceptually on the internationally recognised strategy of sustainable development and the principles of differentiated nature conservation, with each zone having specific environmental and developing functions. In 1993, TANAP was divided into three zones: the core area (49,633 ha, 43.84 %), the buffer zone (23,744 ha, 20.97 %) and the transition area (39,844 ha, 35.19 %). For every functional unit, long-term objectives for the management and conservation of natural resources were specified. The Tatry National Park was the first protected area in Slovakia with official zoning. The 'Management Programme for TANAP' remained in force until 2000. In 1998 a new programme was established on the basis of the new legislation for nature and landscape conservation. However, the proposed new management programme and the new zoning of the park still remain to be approved by government at the time of writing (2010).

Management of the Tatry National Park/Biosphere Reserve

Until 1993 when Czechoslovakia was divided into two countries (Czech Republic and Slovak Republic), protected areas ware managed by the Ministry of Culture, with the exception of Tatry National Park, which was managed by the Ministry of Forests and Water Management. In 1990 the ministries were reorganised. Protected areas were placed under the responsibility of the Ministry of the Environment. However, this Ministry is only responsible for environmental policy. The actual management of the land in the protected areas is the responsibility of the Ministry of Agriculture, through its Forest Section. The Slovak Academy of Sciences is the most important player for the Tatry BR. This institution hosts the Slovak MAB National Committee which officially coordinates the UNESCO Programme in the country. However, the Tatry BR does not have a single employee or a permanent secretariat or a budget to ensure its functioning. There are no institutional mechanisms for promoting transboundary cooperation between the two national parks in Slovakia and Poland. Naturally, the transboundary biosphere reserve will not function without a joint structure devoted to its coordination.

The agency responsible for nature conservation in the Tatry BR is the administration centre of the National Park, whose headquarter is located in the village of Tatranská Štrba. It is subordinated to the State Nature Conservancy in Banská Bystrica which is subordinated to the Ministry of Environment in Bratislava. As the National Park administration is not a legal entity, construction activities, forestry and agriculture, building of health resorts, and related local activities are controlled by the State Nature Conservancy in Banská Bystrica. Most land-use decisions in the Tatry Mountains are made either by local government or by the Ministry of the Interior offices at national and district level respectively. The state forests in the territory of Tatry National Park (app. 55% is state forests, 45% are non-state forests) are managed by local forest service units supervised by the Ministry of Agriculture.

The cooperation between the headquarters of the Tatry National Park and non-governmental organisations is of high importance, especially between those departments dealing with environmental education and nature conservation management. The administration has begun communicating with associations of towns, villages and landowners respectively with a view to harmonising nature conservation and land use. The activities in the National Park follow annual plans, based on long-term objectives. The action plan for the Tatry BR is identical with the annual working plan for the National Park.

Research and education

Multidisciplinary and detailed research in TANAP started after the establishment of a research station in 1953 in Tatranská Lomnica. It is one of a few scientific field stations in the Slovakian mountains and provides a base for research throughout the Tatras. To provide a scientific basis for the rehabilitation of degraded ecosystems, detailed analyses of all environmental components have been undertaken by scientists of the TANAP Research Station, together with scientists of other research institutions of Slovakia. The principal task of the scientists of the TANAP Research Station and their collaborators is the environmental monitoring in the biosphere reserve, with regard to long-term environmental impacts deriving from both the high number of visitors and from air and water pollution.

Environmental education in the Tatry BR includes the development of activities with regard to special educational facilities (e.g. the TANAP Museum in Tatranská Lomnica, information centres or visitor centres for nature conservation, an exhibition of high-mountain flora in Tatranská Lomnica), several educational activities (Educational Natural Trail), as well as publication and promotional activities. Within the educational activities it will be necessary to set a standard for guiding activities on the TANAP territory, to cooperate with cultural and educational institutes, schools, accommodation providers, the press, radio and television, and with other organisations both within the country and abroad.

Historical and actual land use in Tatry Mountains

Mining activities in the High and Western Tatry Mountains started in the 15th century. The longest and most intensive exploration was done on the Kriváň peak. The highest galleries were drilled at an altitude of up to 2,100 metres but yielded only miniscule amounts of the expected gold. The fruitless operation was closed down in 1787. Almost simultaneously with the mining exploration described above, iron mining started in some valleys, along with copper, silver and gold mining in the main ridge of the Western Tatry Mountains. After 1871, open-pit mines (sand-pits and quarries) appeared in connection with the construction of roads, railroads, hotels and sanatoria in the 'Tatranské Foothills'. The largest impact on the natural environment was caused by the sand-pit near the village of Tatranská Polianka and a quarry near Tatranská Kotlina, which the Tatry National Park Administration managed to close in 1958 and 1961 respectively. All historic subsurface mining operations were abandoned before the establishment of the Tatry National Park in 1949. Forests were cut down and converted to charcoal which was needed in smelting works and foundries.

Agriculture is controlled by climate, terrain and soil conditions, which only allow the production of resistant and relatively undemanding cereals. In the past these were mainly hemp and flax. Since the 18th century, the principle crop has been potatoes. Vegetable and fruit trees are cultivated only in small gardens. Since the beginning of its historical development, farming involved cattle and sheep breeding, for which – since the Middle Ages – alpine pastures were used, mainly in the Belianske and Western Tatry Mountains. Grazing areas extended locally over 2,000 metres, especially at the peak of the expansion of shepherding in the 17th and 18th centuries. Grazing had a number of adverse effects and did considerable damage in the area. The destruction of the environment was mainly caused by shepherds, who – in order to obtain more grazing areas – set about felling

Impressions from Tatry National Park and Biosphere Reserve (© Ivan Bohuš).

dwarf pine stands and the uppermost parts of the forests. It was also caused by the impact of animal hooves which degraded and eroded the slopes. Despite all endeavours, the rehabilitation of the original habitat has not yet been completely successful. Cattle and sheep breeding partially decreased in the second half of the 19th century, mainly as a result of the region being industrialised. The grazed area in alpine pasture was gradually reduced by land reclamation of more accessible and higher-quality sub-montane pastures and meadows. Sporadic attempts at grazing continued even after the establishment of the Tatry National Park, but after 1955 they were successfully eliminated for the benefit of nature conservation.

Since the end of the 19th century, irresponsible forest management gave rise to calls for an expropriation act, but such legislation was never passed. The Hungarian Ministry of Agriculture tried to solve this problem by systematic acquisition of the most threatened forest land, which shortly before had started to be purchased by wealthy foreigners, mainly by the Duke of Hohen-lohe and by Baron Diergardt. After 1918, the new Czechoslovak Republic persisted with this approach which culminated in the wholesale purchase of the Tatry land from the two landowners mentioned above. At the time when Tatry National Park came into being, 29,331 hectares of the land, out of the total territory of 43,505 hectares, was in state ownership, and by 1958 another 18,827 hectares became the outright property of the former Tatry National Park Administration. In 1987, the Western Tatry Mountains were integrated into the Tatry National Park. Claims for unlawful interference by the totalitarian regime in private ownership are currently being addressed by restitution of procedural standards. By 1st January 1993, a total of 29,875 hectares had been returned to the competent owners. However, territorially, these estates remain part of the Tatry National Park, and remain subject to officially authorised forest-management plans for their management. Nowadays, the state owns 54 per cent of the total area of Tatry National Park. Some 65 per cent of the total forest area is classified as protective forest (ecological functions: soil and water conservation), 35 percent are subject to the category of special objectives (environmental functions: recreation, health, spa, nature conservation, etc.).

Tourism, as we know it today, started to appear in 1871, when the Podtatranská Basin was linked with the rest of the world by the Košice-Bohumín railway line. Hotels were built close to its stations, and in the high-altitude zone a number of climate centres were developed, first in Starý Smokovec, established as early as 1793, and then all the way from Štrbské Pleso to the Tatranská Kotlina settlement. Some of them were a kind of climate spa, but eight out of 16 are today mainly tourist centres whereas another three have both functions. Before the First World War, the number of visitors in the Tatry Mountains was approx. 10,000 per year; in the period between the Wars the figure peaked at 25,000 per year, by 1970 it increased to 500,000 per year and today (2009) it is an unmanageable figure of approx. three million visitors annually. The level of service in the Tatry settlements kept adapting to fashion trends and the changing requirements of domestic and foreign clientele. Until 1885, the operation of tourist establishments was limited to the summer period, but later, summer and winter tourism became the norm. This fact was overlooked when new tourist facilities were built, and therefore the Tatry Mountains are now characterised by over-capacity which is not used in the periods between seasons.

Mountaineering became a sport in the beginning of the 20th century and reached great popularity after 1949 when crampons came into wider use. The first international sledging competitions were held in 1903 and a toboggan run was built in Tatranská Lomnica in line with European parameters. Since 1905, both Alpine and Nordic skiing became popular and in 1911 Tatranská Polianka hosted the first international competitions in Alpine skiing. In 1935 and 1970 the World Championships in Nordic skiing were held in the Mlynická Dolina valley, where a modern sport resort for Nordic disciplines was built. The Slovak Grand Prix competitions in Alpine skiing were also held here (since 1967 alongside the World Cup competitions). This territory also has other well-equipped ski resorts, e.g. in Hrebienok, Solisko, Tatranská Lomnica, Skalnaté Pleso, and – for recreational skiing – in Ždiar. For ecological reasons, further ski-resort development was kept out of valuable landscape environments (e.g. the Slavkovská Dolina valley and Adamcula in the Roháče Mountains).

Unfortunately, the development of the biosphere reserve was not restricted to the transition area, but intruded into the core zone. Already the plan of territorial development in the High Tatry Mountains from 1959 included the reversal of the park's priorities: Recreation, tourism and sports, heath care and research. From 1964 onwards, this functional hierarchy was also applied to the concept of the Tatry National Park. The subsequent chaotic development brought an overload in terms of visitors, culminating in the period from 1980 to 1982. Owing to a lack of economic opportunities, this development was not sufficiently recognised, especially in the spheres of transport and sewage treatment.

Apart from direct negative impacts, for example on the scenic beauty of the landscape destruction of wildlife areas (e.g. more than 100 hectares of forest were taken up in the Štrbské Pleso area by the World Championship in Nordic skiing in 1970) and damage caused by motor vehicles, there was also damage to ecosystems in the forest zone and in the alpine environment. This damage was first manifest in the vicinity of the Tatry settlements: camping sites, chalet areas, transport facilities, mountain huts and tourist trails. In addition to the destruction of ecosystems, contamination with refuse, and damage to the vegetation, another serious problem was the unsuitable behaviour of visitors with regard to wild animals, especially their intentional disturbance by photographing, filming and feeding. Similar damage is caused by organised tourist activities.

In the alpine environment these are mainly mountaineering, ski-alpinism, hang-gliding, parachuting, and latterly also mountain biking.

In spite of the fact that after 1989, the number of visitors decreased by about 40 to 50 per cent, there are still localities with a surplus of visitors (e.g. Skalnaté Pleso and Solisko), where the TANAP Administration implements a number of technical measures in order to limit damage to the park's environment. The single most serious adverse effect on Tatry wildlife occurred in 2008 when ski sport facilities were built in Tatranská Lomnica – the Skalnaté Pleso and Solisko localities. Meanwhile, the International Union for Conservation of Nature (IUCN) stated that the developments taking place in the Tatry National Park are not in line with the objectives of the Category II classification. Currently, the downgrading to a lower category, for instance V, is being discussed by the IUCN. Owing to the changes in land use and damage to ecosystems in the former core zone, it is very urgent to establish a new zonation for the Tatry BR. However, the current process of re-zoning Tatry National Park incurred the protest of environmental groups and scientists. They are critical of the proposal that intensive development be allowed even in the most sensitive parts of the area. More than 472 hectares of formerly protected highland meadows and forests are considered to be eligible for tourism facilities. The opponents fear that the most important breeding sites of chamois, marmot, black grouse and capercaillie in the West Tatry Mountains would be destroyed by those plans.

References

Vološčuk, I. (ed.) (1994). The Tatras National Park – Biosphere Reserve (in Slovak). Gradus Martin.

Worldwide Case Studies

The Role of Biosphere Reserves in Research and Monitoring

Learning from experiences

Chapter 3-1

Monitoring global change in mountain regions, e.g. in Cairngorms, Scotland (© Harald Pauli).

Monitoring Global Change in Mountain Biosphere Reserves: GLOCHAMORE, GLOCHAMOST and GLORIA

by Sigrun Lange

The Madrid Action Plan identified climate change as one of the *'most serious and globally significant challenges to society and ecosystems around the world today. The role of biosphere reserves is essential to rapidly seek and test solutions to the challenges of climate change as well as monitor the changes as part of a global network'* (UNESCO MAB 2008, p.6).

'GLOCHAMORE' – a strategy to monitor global change in mountain regions

Already in 2003, by launching the GLOCHAMORE project (Global Change and Mountain Regions), the Mountain Research Initiative (MRI) and the University of Vienna (Austria), in collaboration with UNESCO's MAB Programme, responded to the increasing need to understand the causes and impacts of global changes in mountain regions. In the course of the project, funded within the EU's Sixth Framework Programme, 25 biosphere reserves in mountain regions all over the world have been chosen as pilot regions. With their gradient from little human disturbance in the strictly protected core zones to populated and strongly developed areas in the transition zones, biosphere reserves (BRs) were considered interesting study and monitoring sites to assess global change impacts in mountain ranges. The main outcome of the project was the GLOCHAMORE Research Strategy (MRI 2005) released after five thematic workshops and a final open science conference 2005 in Perth, Scotland. It is organised by themes, starting with drivers of global change, continuing with the impacts of global change on ecosystems, their goods and services and people's well-being, and closing with themes related to adaptation measures. The knowledge from both (natural and social) science and from UNESCO mountain BR managers has been incorporated in the strategy. It has been developed to guide managers of mountain BRs and scientists in planning and implementing their research and monitoring activities. The corresponding scientific results are intended to serve as a basis for BR managers and other stakeholders to develop sustainable development policies for their respective sites. With the publication of the GLOCHAMORE Research Strategy, the EU project has been concluded. In the 'Perth Declaration' site managers and scientists declared that they would strengthen further the global change research in the selected biosphere reserves in accordance with the outlines of the strategy. Five years later, in September 2010, the GLOCHAMORE key players and other scientists from across the globe met again in Perth to take part in the largest-ever conference on 'Global Change and the World's Mountains' organised by the Perth College UHI's Centre for Mountain Studies (CMS) and the global Mountain Research Initiative (MRI). Nearly 500 scientists from 60 countries presented their results from surveys in mountainous areas. According to Martin Price who opened the conference, the meeting provided a renewed focus for mountain issues and global change. The results will facilitate the process of drafting an action plan for mountains in the run-up to the 'Rio + 20' United Nations Conference on Sustainable Development in 2012 (Perth College UHI 2010).

'GLOCHAMOST' – towards the implementation of the GLOCHAMORE Research Strategy

Since 2005, BR managers and scientists have been encouraged to implement all aspects of the GLOCHAMORE Research Strategy (in line with their own needs and priorities), but in fact, most of them may not be able to afford the necessary human resources and technical infrastructure needed for such a comprehensive undertaking (Schaaf 2008). In the discussion process, four key research areas have proved to be of particular importance (UNESCO MAB 2010), i.e. the

- Impact of global change on key fauna and flora (item 6e of the GLOCHAMORE Research Strategy);
- Availability of freshwater resources in the context of global warming (item 4a of the Research Strategy);
- Understanding the origins and impacts of land-use changes (item 2b of the Research Strategy);
- Development of mountain economies and livelihoods of mountain dwellers (item 9a of the Research Strategy).

UNESCO MAB's new project 'GLOCHAMOST' (Global Change in Mountain Sites) now aims at implementing these specific elements of the GLOCHAMORE Research Strategy in representative mountain BRs, with a view to developing adaptation strategies which address the specific impacts of global change on these environments, their inhabitants, and others who depend on goods and services deriving from these mountain areas (Schaaf 2008). Several BRs are already studying the key research areas mentioned, such as Katunskiy (cf. p. 57) and Teberdinskiy BRs (both Russian Federation), Sierra Nevada BR (Spain, cf. p. 52), Changbaishan BR (China), Nanda Devi BR (India, cf. p. 79) and Huascaran BR (Peru).

'GLORIA' – Observing the impact of global change on the high-mountain flora

For ten years, the 'Global Observation Research Initiative in Alpine environments' (GLORIA) has endeavoured to monitor global change effects on vegetation in high-mountain environments. The demand for a comparative observation network in mountain ecosystems was already highlighted in 1996 during a workshop in Kathmandu (Nepal). Subsequently, Georg Grabherr from the University of Vienna (Austria) and his colleagues developed a first concept for such a network which, in 2000, was presented to an international audience: 'GLORIA' was launched (GLORIA 2010). Meanwhile, the network consists of permanent observation sites in more than 75 mountain regions on five continents, involving about 60 working groups (Grabherr, Gottfried & Pauli 2010). Five international meetings have already taken place, the last one in September 2010 in Perth, Scotland. The so-called 'Multi-Summit Approach' is based on the periodic survey (every 5 to 10 years) of plant communities across summits at four different elevations, representative of a particular mountain region; a treeline summit; a summit at the transition from low to high alpine; one reaching to the alpine-nival ecotone;

and one to the uppermost limits of plant life. Temperature loggers are inserted into the soil to obtain a time series of temperatures. Snow cover duration can be derived from these measurements (Grabherr, Gottfried & Pauli 2010). The success of the spreading network may be explained by the fact that the establishment of permanent observation plots is simple and cheap. Applying a standardised methodology allows for a comparison of the regions, and a regional to global assessment of how climate change affects high-mountain species (Pauli et al. 2009). Amongst the partners in the GLORIA network is a wide range of protected areas, mainly national parks and BRs from all over the world (cf. case study on Katunskiy BR, p. 57).

Harald Pauli (right), GLORIA coordinator, and a team from the Ecological Institute of Jaca are establishing a new GLORIA site in Ordesa Biosphere Reserve in Spain (© Monika Wenzl).

References

GLORIA (2010). A brief history of GLORIA; GLORIA web site: http://www.gloria.ac.at/?a=11 (accessed on 22 October 2010).

Grabherr, G., Gottfried, M. & Pauli, H. (2010). Climate Change Impacts in Alpine Environments. In: Geography Compass 4/8 (2010): 1133–1153.

MRI (ed.) (2005). GLOCHAMORE – Global Change and Mountain Regions Research Strategy. A joint project of the Mountain Research Initiative (MRI), UNESCO-MAB and IHP, and the 6th EU Framework Programme; available at: http://mri.scnatweb.ch/projects/glochamore/print-version-of-the-glochamore-research-strategy.html (accessed on 22 October 2010).

Pauli, H., Gottfried, M., Klettner, Ch., Laimer, S. and Grabherr, G. (2009). A global long-term observation system for mountain biodiversity: lessons learned and upcoming challenges. In: Sharma, E., (ed.) Proceedings of the International Mountain Biodiversity Conference. Kathmandu: ICIMOD, pp. 120–128.

Perth College UHI (2010). Centre for Mountain Studies Conference. News article from September 29 2010; available at http://www.perth.ac.uk/news/Pages/CentreforMountainStudiesConference.aspx (accessed on 22 October 2010).

Schaaf, T. (2008). Global Change in Mountain Regions – Strategies for Biosphere Reserves. Contribution to the Plenary Session on 'Climate Change and its Implications for Mountains' of the International Mountain Biodiversity Conference from 16 –18 November 2008 in Kathmandu, Nepal. Available at www.icimod.org/resource.php?id=48 (accessed on 22 October 2010).

UNESCO MAB (2010). GLOCHAMOST (Global and Climate Change in Mountain Sites – Coping Strategies for Mountain Biosphere Reserves); information at UNESCO's web portal: http://portal.unesco.org/science/en/ev.php-URL_ID=8991&URL_DO=DO_TOPIC&URL_SECTION=201.html (accessed on 22 October 2010).

UNESCO MAB (2008). Madrid Action Plan for Biosphere Reserves (2008 – 2013).

Mountain scenery in Sierra Nevada National Park and Biosphere Reserve (© José Miguel Muñoz).

Sierra Nevada Observatory for Monitoring Global Change: Towards the Adaptive Management of Natural Resources

by F.J. Bonet, R. Aspizua, R. Zamora, F. Javier Sánchez, F. Javier Cano-Manuel & I. Henares

The Sierra Nevada (Spain) was declared a Biosphere Reserve (BR) by UNESCO in 1986. In 1999, 85,883 hectares in the centre were declared a national park incorporating the Mediterranean high-mountain ecosystems. The massif includes Mulhacén – at 3,482 metres the highest summit in the Iberian peninsula. The Sierra Nevada mountain range is one of the most important hotspots of biological diversity and endemicity in the Iberian Peninsula and is therefore an exceptional observatory for studying the functioning of natural systems and processes under the current global-change scenario. The 'Sierra Nevada Observatory for Monitoring Global Change' has a long-term vocation and seeks permanent cooperation between scientists and managers. Its objective is to obtain information that helps identify the impacts of global change as early as possible, enabling the design of management mechanisms to minimise its impacts. It has four cornerstones: a programme for monitoring key species, ecosystems and processes, an information system that translates data generated by the programme into useful knowledge, an active adaptive management of natural resources which also serves to provide results and feedback to feed into this knowledge base; it provides some efficient tools for the continuous training of managers, and, finally, it provides effective dissemination mechanisms that inform society of results obtained and methodologies used, enabling comparisons with other experiences made elsewhere in the world.

Project background and origin

The Sierra Nevada Observatory for Monitoring Global Change stems from an international initiative sponsored by UNESCO and the Sixth Framework Programme of the European Union, called GLOCHAMORE (Global Change in Mountain Regions). This project was initiated in 2003, with the objective of developing protocols that monitor the effects of global change in mountain regions. It combines long-term monitoring with monitoring at a global scale, taking into account the human dimension and biotic resources, while seeking continuous cooperation between scientists and managers. The GLOCHA-MORE project proposes a joint research strategy between scientists and managers of the protected areas affected (Björnsen

et al. 2005). This document was the starting point of the current Sierra Nevada Monitoring Observatory. Participation in the GLOCHAMORE project revealed something that has made a significant contribution to determining the structure and operation of the Sierra Nevada Observatory: It is vital to tackle the study of global change in a combined approach to scientific research and the sustainable and active management of natural resources. In other words, it is essential that the monitoring protocols introduced are scientifically approved and provide the managers of individual territories with useful decision-making information. This is to make sure that, thanks to a philosophy of continuous adaptation to change, the decisions made can in turn lead to actions which may help provide information that feeds back into the process.

Currently, and as a continuation of GLOCHAMORE, the MAB Programme is promoting the 'Global Change in Mountain Sites (GLOCHAMOST)' project to implement the previous GLOCHAMORE strategy which promotes cooperation and communication between industrialised and developing countries, and facilitates collaboration between researchers, mountain biosphere reserve managers and the communities affected by global change (Schaaf 2009). Only ten biosphere reserves are involved in designing this initiative, because – along with the Sierra Nevada BR – other areas were selected in countries such as China, Germany, India, Peru, Russia, Switzerland and the United States.

Objectives and cornerstones

The Sierra Nevada Observatory for monitoring global change intends to obtain the information necessary for identifying as early as possible the impacts of global change, in order to design management mechanisms that help minimise these impacts and adapt the system to new situations. To this end, the following general objectives are considered:

- Evaluating the working of ecosystems in the Sierra Nevada Nature Reserve, their natural processes and dynamics over a medium-term timescale.
- Identifying population dynamics, phenological changes

and problems with the conservation of key organisms as indicators of ecological processes that might be affected by climate change.

- Identifying the possible effects of global change on monitored species, ecosystems and natural resources, providing an overview of trends of change that help to create adaptability throughout the ecosystem.
- Designing mechanisms for evaluating the effectiveness and efficiency of management activities carried out in the Sierra Nevada in order to propose appropriate adjustments to implement an adaptive management model.
- Providing basic information for periodic planning tasks in protected areas.
- Helping to disseminate information of general interest that enhances knowledge of the values and importance of the Sierra Nevada.

The achievement of these objectives firstly requires designing and implementing a programme that monitors the effects of global change on the Sierra Nevada. That is why it is vital that all data collected is integrated and analysed in the context of an information system associated with the project. Finally, it is important to inform society of both the results and work methodologies adopted, through effective dissemination mechanisms. The monitoring programme produces data which is processed by the information system to generate useful knowledge for the management of resources. For easy reference, the standardised data is entered into freely accessible databases. The four cornerstones of the monitoring programme (monitoring programme, adaptive management, information systems and dissemination) are described below in more detail.

Monitoring programme

The design of global change adaptation mechanisms requires information on the structure and dynamics of different ecological elements and processes. The design of this monitoring programme is based on the thematic areas of the GLOCHA-MORE project. Different monitoring methodologies were defined for each of these thematic areas to assess both the state of key ecological functions and the structure of the main ecosystems in the Sierra Nevada. A solid monitoring programme should allow its users to be aware of the past, in order to understand the present and try to adapt to the future. This is particularly important in Mediterranean ecosystems which have been managed by humans for centuries. In terms of the distribution of monitoring operations in the area, the design has taken into account the enormous heterogeneity and diversity of the mountain range. The sampling units are therefore distributed throughout the five bioclimatic zones in the Sierra Nevada, in the different types of existing ecosystems, covering maximum environmental diversity.

Management of information generated

In parallel with compiling information on the state and structure of the Sierra Nevada's natural systems, a database is being prepared for storing all this information. This tool will supply managers with useful information (i.e. knowledge) for improving the way in which the Sierra Nevada's natural resources are managed. This knowledge is obtained once the raw data obtained by the monitoring programme described above has been processed and analysed. The information system is based on the design of procedures ensuring the orderly storing of information generated by the monitoring programme. These databases are documented by means of metadata standards used by the LTER Network (Fegraus et al. 2005). The algorithms used in analysing and processing the above data are also automatically documented and executed by means of workflow management applications (Barseghian et al. 2010).

Adaptive management

In contrast with traditional management lacking in monitoring, it is necessary to strengthen management and adaptive monitoring (Lindenmayer & Likens, 2009), demonstrating the value of protected areas as natural laboratories for testing new techniques for managing and monitoring natural resources that improve the day-to-day management practice. Once validated, these new techniques can be exported for immediate application to the rest of the territorial matrix. The active management of ecosystems in a global-change scenario requires the adoption of a flexible management approach because current forms of management might not be applicable in the future. That is why it is now necessary more than ever to implement projects that assess the suitability of new and old management techniques in the light of global change scenarios. The basic elements of active adaptive management can be summarised as follows:

- Objectives outlining management of natural resources and hypotheses for their achievement (including monitoring indicators).
- Immediate recording of data (monitoring indicators).
- Assessment of progress of monitoring results. Improvement in knowledge of natural processes that govern the management of ecological systems.
- Adjustment of activities and natural-resources management policies through changes in line with results obtained and lessons learned. This may also mean an adjustment of resources.
- Documentation of process and results.
- Dissemination of knowledge gained and provision of access to experts with regard to specific data.

Fig. 1: Monitoring activities in Sierra Nevada Biosphere Reserve (© José Miguel Muñoz).

Examples for adaptive management are large-scale projects for conservation and the improvement of oak and juniper woodlands for better adaptation to the impacts of global change. These plant communities are being degraded by environmental conditions and changes in land use. To improve the resilience of these types of ecosystems and favour their regeneration in view of changing conditions in their high-mountain habitats, selective forest clearing and sanitation activity is carried out to reduce competition, as well as sowing and planting to encourage the persistence of oak and juniper species in areas which, according to predictions made by the climate models produced for the Sierra Nevada (Benito de Pando, 2009), will be more favourable in the next few decades.

Another example is the experimental treatment applied after the fire at Lanjaron in September 2005, in which more than 3,000 hectares, mostly pine reforestation, burnt down. One of the issues raised after the fire was to determine the role of deadwood in natural regeneration after this type of event. The reason for this is obvious: Although traditional restoration includes cutting and removing burnt logs, along with the chipping of branches, there are very few studies evaluating the general suitability of this procedure. It is important to know whether these actions affect the recruitment of seedlings and new shoots, and in that case to what extent, as well as their effect on diversity and nutrient recycling. With the objective to learn more about the handling of wood burnt in this type of process thus deepening the knowledge with respect to the most suitable restoration in line with the peculiarities of the environment, a large-scale experiment was staged to study the response of the ecosystem to three different treatments: (a) the traditional cutting and removing of wood combined with chipping of branches, (b) cutting the majority of trees, leaving branches piled up covering approx. one third of the soil surface and trees left upstanding to function as perches for seed-scattering animals, and (c) rather than cutting or extracting, to leave burnt trees standing. Nine plots were marked (Figure 2 shows an aerial image), with three replicates for each treatment, occupying the total experimental surface of 112 hectares that were excluded from the traditional management carried out in the rest of the burnt area. The impact of the different management options on plant regeneration ability was studied in terms of seeding and regrowth, along with the success of artificial plants getting established, and the diversity of the community. Furthermore, several biotic and abiotic parameters have been quantified, such as nutrients and water availability, radiation, soil compaction, and wood decomposition rates. This has allowed the development of restoration models adjusted to the environmental heterogeneity, including the natural recolonisation of oak (*Quercus ilex subsp. Ballota*) with a different recruitment rate observed in the different treatments studied. Finally, the economic costs of both implementation and maintenance of the experimental handling are being assessed, which will allow the inclusion of ecological, economic and management criteria for the development of the most appropriate restoration models.

Fig. 2: Aerial image showing the distribution of post-fire regeneration plots in Lanjaron.

Dissemination

For dissemination to be successful, it is important to first define the 'targets' for disseminating the results. In our case they are as follows:

- General public: society must be made aware of the importance of the Sierra Nevada as a key ecosystems services supplier (for example, water supply) for most of Andalusia.
- Managers: in line with the project's philosophy, the area's managers must be the main recipients of results. The basic idea is for the Sierra Nevada Observatory to supply useful knowledge for improving the way in which managers take decisions on managing natural resources.
- Scientists: the ecological uniqueness of the Sierra Nevada makes it a major focus of interest for scientists from different disciplines, allowing access to the raw data generated by the monitoring programme.

Bearing in mind these three main recipients of the Observatory's results, we can identify different types of dissemination measures:

- A set of procedures designed to communicate the Observatory's results to the general public.
- Training for the technical experts from the protected area management teams with the aim of updating their scientific-technical knowledge, and for researchers, managers and technical personnel in order to exchange experiences and knowledge and share problems regarding the implementation of the project.
- A web portal for consulting and downloading all information compiled by the Sierra Nevada Observatory.
- A project 'wiki' as main communication tool, which allows the collaborative editing of the content; an exchange with other similar monitoring programmes can also be included.

Link with other monitoring networks

The Sierra Nevada Observatory is the consequence of the convergence of two interests: firstly, managers and scientists that carry out their work in the Sierra Nevada have shown their interest in the Observatory which, as we have described, is becoming a reality. Secondly, there are other initiatives arising from international institutions (GLOCHAMORE) which have promoted the implementation of this project. The result of these two trends (bottom-up and top-down) is the creation of a project for a specific territory, applying methodologies compatible with other areas, and characterised by a strong motivation to collaborate with other similar initiatives.

This commitment is demonstrated by the involvement of the Sierra Nevada Observatory in the design and implementation of similar projects both regionally and nationally. We are also helping to create and consolidate the Global Change Observatories Network in Andalusia, the region where the Sierra Nevada BR is located. Nationally, the Sierra Nevada Observatory forms part of the LTER Network in Spain, and is involved in the monitoring programme sponsored by the Spanish National Park Service.

Acknowledgements

The authors would like to thank the Department of Environment of the Regional Government of Andalusia for its continuous support through various financing methods as mentioned above, as well as the Ministry of the Rural, Marine and Natural Environment for the essential provision of scientific instruments through the Global Change in National Parks Monitoring Network. Thanks also to the University of Granada for its important scientific collaboration through the Andalusian Environment Centre and, finally, for the enthusiastic support by personnel from the Sierra Nevada Nature Reserve and the public company of EGMASA.

References

Aspizua Cantón, R, Cano Manuel-León, FJ, Bonet García, FJ, Zamora Rodríguez, R & Sánchez Gutiérrez, J (2007). Sierra Nevada: Observatorio internacional de seguimiento del cambio global. (Sierra Nevada: International observatory monitoring global change.) In: Revista Medio Ambiente, 57: 21–25.

Barseghian DI, Altintas, MB, Jones, D, Crawl, N, Potter, J, Gallagher, P, Cornillon, M, Schildhauer, ET, Borer, EW, Seabloom & Hosseini, PR (2010). Workflows and extensions to the Kepler scientific workflow system to support environmental sensor data access and analysis. In: Ecological Informatics, 5: 42–50.

Benito de Pando, B (2009). Ecoinformática aplicada a la conservación: Simulación de efectos del cambio global en la distribución de la flora de Andalucía. Doctoral thesis, University of Granada.

Björnsen A, Becker, A, Brun, J, Bugmann, H, Dedieu, J, Grabherr, G & Haeberli, W (2005). GLOCHAMORE, Global Change and Mountain Regions. Research Strategy. UNESCO Man and the Biosphere (MAB) Programme, and the UNESCO International Hydrological Programme (IHP). Ed. GLOCHAMORE Scientific Project Manager.

Blanca López, G, López Onieva, MR, Lorite, J, Martínez Lirola, MJ, Molero Mesa, J, Quintas, S, Ruíz Girela, M, Varo, MA & Vidal, S (2001). Flora amenazada y endémica de Sierra Nevada. Department of the Environment of the Regional Government of Andalusia and the University of Granada.

Bonet García, FJ, Villegas Sánchez, I, Navarro, J & Zamora, R (2009). Breve historia de la gestión de los pinares de repoblación en Sierra Nevada. Una aproximación desde la ecología de la regeneración. In: Proceedings of the 5th Spanish Forestry Congress.

Bonet García, F.J. & Cayuela Delgado, L. (2009). Seguimiento de la cubierta de nieve en Sierra Nevada: tendencias en la última década y posibles implicaciones ecológicas de las mismas. In 11th National Congress of the Spanish Association of Terrestrial Ecology: The ecological dimension of sustainable development: Ecology, from knowledge to application. Úbeda, 18–22 October 2009.

Borrini-Feyerabend, G., Farvar, M. T., Nguinguiri, J. C. & Ndangang, V. A., (2007). Comanagement of Natural Resources: Organising, Negotiating and Learning-by-Doing. GTZ and IUCN, Kasparek Verlag, Heidelberg (Germany).

Castro, J., Navarro-Cerrillo, R., Guzmán-Álvarez, J. R., Zamora, R. & Bautista, S., (2009). ¿Es conveniente retirar la madera quemada tras un incendio forestal? (Is it is advisable to remove burnt wood after a forest fire?) Quercus 282: 38–43.

Castro, J. , Allen, C. D., Molina-Morales, M., Marañón, S., Sánchez, Á. & Zamora, R., (2010). Salvage Logging Versus the Use of Burnt Wood as a Nurse Object to Promote Post-Fire Tree Seedling Establishment. Restoration Ecology (in press).

Duarte, C. (Coord.) (2006). Cambio Global: Impacto de la actividad humana sobre el sistema Tierra. (Global change: Impact of human activity on the Earth system.) 167 pp. Madrid; CSIC (Spanish National Research Council).

Fegraus, E. H. , Andelman, S., Jones, M. B. and Schildhauer, M. (2005). Maximizing the Value of Ecological Data with Structured Metadata: An Introduction to Ecological Metadata Language (EML) and Principles for Metadata Creation. Bulletin of Ecological Society of America. Vol. 86 pp.158–168.

Holling, C.S. (1978). Adaptive Environmental Assessment and Management. 377 pp. John Wiley & Sons., New York.

IPCC, (2007). Informe de síntesis. Contribución de los Grupos de trabajo I, II y III al Cuarto Informe de evaluación del Grupo Intergubernamental de Expertos sobre el Cambio Climático. (2007 climate change: Summary report. Contribution of Working Groups I, II and III to the Fourth Assessment Report of the Intergovernmental Panel on Climate Change.) 104 pp. IPCC, Geneva, Switzerland.

Lee, K. N. (1999). Appraising adaptive management. Conservation Ecology 3(2): 3 (available at http://www.consecol.org/vol3/iss2/art3/).

Navarro González, I., Bonet García, F.J. (2009). Caracterización de la evolución histórica de la cubierta vegetal y los usos del suelo de Sierra Nevada en un contexto de cambio global. At 11th National Congress of the Spanish Association of Terrestrial Ecology: The ecological dimension of sustainable development: Ecology, from knowledge to application.

Nyberg, J.B. (1998). Statistics and the practice of adaptive management. Pages 1–7 in Statistical Methods for Adaptive Management Studies, V. Sit and B. Taylor, (editors). Land Manage. Handbook 42, B.C. Ministry of Forests, Victoria, BC.

Nyberg, B. (1999). An Introductory Guide to Adaptive Management. Forest Practices Branch. 24 pp. B.C. Forest Service (Canada).

Pérez-Luque A. J., F. J. Bonet García, and R. Zamora Rodríguez. (2009). Herramientas colaborativas para la creación de conocimiento útil para la gestión en el proyecto de Seguimiento del Cambio Global en Sierra Nevada. At 11th National Congress of the Spanish Association of Terrestrial Ecology: The ecological dimension of sustainable development: Ecology, from knowledge to application.

Ruano, F. and Tinaut, A. 2003. Historia de la entomología en Sierra Nevada (Sur de España) de 1813 a 2000. Boln. Asoc. esp. Ent., 27 (1–4):109–126.

Sánchez-Gutiérrez, F.J.; Henares-Civantos, I.; Cano-Manuel León, F.J.; Zamora Rodríguez, R.; Bonet García, F.J. & Aspizua Canton, R. (2009). El observatorio de cambio global de Sierra Nevada. (The Sierra Nevada global change observatory.) Revista Medio Ambiente, 63: 16–19.

Schaaf, T. (2009). Mountain Biosphere Reserves – A People Centred Approach that also Links Global Knowledge. Sustainable Mountain Development No. 55, International Centre for Integrated Mountain Development (ICIMOD), Spring 2009.

Speth, James G, 'The Global Environmental Agenda: Origins and Prospects' in Daniel C. Esty and Maria H. Ivanova, eds., Global Environmental Governance: Options and Opportunities. New Haven, CT: Yale School of Forestry and Environmental Studies, 2002.

Vitousek, Peter M., Harold A. Mooney, Jane Lubchenco, and Jerry M. Melillo. (1997). Human Domination of Earth's Ecosystems. Science 277 (5325): 494–499.

Woolman M. (2002). Digital Information Graphics. Watson-Guptill eds.

Zamora, R. & Sánchez, F.J. (2009): El valor de los Espacios Protegidos de montaña en un escenario de cambio global. Ecosistemas, 18 (3): 35–37.

GLORIA target region in the Katunskiy BR (© Igor Artemov).

Monitoring Climate Change Effects in the Katunskiy Biosphere Reserve (Russian Federation)

by Tatjana Yashina & Igor Artemov

According to the vision of biosphere reserves, its functions should not only be limited to conservation and fostering sustainable regional development. Core zones of biosphere reserves are excellent natural laboratories; they are not transformed by human activities, their ecosystems are represented in their natural, undisturbed state. The Madrid Action Plan advises to 'use biosphere reserves (BRs) as learning sites for research, adaptation, mitigation in relation to climate change' and furthermore 'to use mountain BRs as field observatories of global change impacts on the environment, economy and human well-being, based on the GLOCHAMORE Research Strategy' (UNESCO-MAB 2008, p. 24).

The Katunskiy BR is located within the Ecoregion of Altai-Sayan, designated as one of the WWF Global-200 Ecoregions of the World. Located in the Central Altai Mountains, Katunskiy BR covers more than 600,000 hectares, including 151,600 hectares of intact high-altitude landscapes as the core zone. The land cover of the Biosphere Reserve is represented by the following types of landscapes: glaciers and nival landscapes (covering 24% of its surface area), alpine landscapes (30%), boreal mountain forests (43%), a combination of woods and steppes (2%) and steppes in intermountain depressions (1%). The highest peak of Siberia, Mt. Belukha (4,506 m.a.s.l.), is located within the transition zone of the park. The Altai Mountains and the biosphere reserve territory, which provide a significant source of fresh water conserved in glaciers, could be considered as a water tower for the vast expanse of western Siberian lowlands. The estimated amount of water stored in the glaciers is 7.37 trillion tonnes (Galakhov & Mukhammetov 1999).

The biological diversity of the Katunskiy BR consists of approx. 1,000 species of higher vascular plants (including 9 endangered species), 161 species of birds (20 are endangered) and 52 species of mammals (two are endangered). Habitats and important migratory routes of globally endangered species such as the snow leopard (*Uncia uncia*), musk-deer (*Moschus moschiferus*), black vulture (*Aegypius monachus*), fish-hawk (*Pandion haliaetus*) and others, are located within the biosphere reserve, in particular within its core zone. In 1998, Katunskiy BR was designated as the cluster of the 'Golden Mountain of Altai' UNESCO World Heritage Site, nominated for its extremely rich and globally significant biological diversity.

In its location far from large towns, cities and industrial centres, the core zone of the Katunskiy BR contains natural ecosystems that have not been transformed by human activities. The key driver of its dynamics is climate change. The regional trend of mean annual temperature, calculated by the analysis of climatic data from Barnaul weather station, established in 1835, shows significant warming from -0.5 to +2.3°C. Local temperature variations are not so clear, since Central Altai is characterised by the lowest rates of warming in the region. Nevertheless, the observations from meteorological stations located within the Katunskiy BR show the general trend of warming. Kharlamova (2010) provides an analysis of climatic data for a 50-year period. This is data from meteorological stations located at different altitudes which indicate an increase in the mean annual temperature by +1.45°C at high elevations (2,000 m.a.s.l.) and by +2.1°C at lower elevations within intermountain depressions (998 m.a.s.l.) during the last five decades.

Ecosystems at high altitudes are particularly sensitive and vulnerable to climate change. An increase in the air temperature may cause upward shifts of altitudinal belts, significant changes in vegetation and habitat patterns, especially in summit areas. Therefore, studies of natural processes caused by climate change at high elevations (in particular in the alpine zone) are of vital importance for further projections of ecosystem response to changes. The territory of the Katunskiy BR is one of the most significant centres of modern glaciation in Siberia with 317 glaciers which cover a total surface area of 230 square kilometres (Galakhov & Mukhammetov, 1999). Recent studies show that the glaciers of the Altai Mountains have retreated by 19.7 per cent during the period 1952–2004. At the same time, large glaciers of the Belukha massif have lost about 15 per cent of their surface area (Galakhov & Mukhammetov 1999, Narozhny et al. 2006). This accelerated glacier melting causes changes in the hydrology of high-altitudinal catchments, because melt water forms more than 50 per cent of the total discharge of rivers in upper and middle-elevation zones. Many studies demonstrate intensive forest growth and an upward shift of forest vegetation by 30 to 60 metres during the last 60 to 80 years in different regions of the world. Such a trend is also observed in the Katunskiy BR, where the upper tree line has shifted upward by 60 to 100 metres during the last 120 years (Patrusheva 2010).

High-altitude ecosystems, located above the tree line, are formed by abiotic, mainly climate-related factors, while the role of biological factors (such as competition) decreases with altitude. Therefore such ecosystems with temperatures close to the lower limits of plant survival are the most sensitive and vulnerable to climate change. Such alpine ecosystems are the focus of research by GLORIA, Global Observation Research Initiative in Alpine Environments, an initiative which intends to establish an international research network to assess climate change impacts on mountain environments. This initiative provides a standardised sampling design for monitoring alpine environments such as GLORIA's Multi-Summit Approach (Pauli & Gottfried 2004). In 2005, a GLORIA site was established in the Katunskiy BR within the framework of GLOCHAMORE (which is dedicated to the detection of changes in alpine ecosystems), with support from the GLORIA coordination team (Vienna University) and the UNESCO MAB Programme. For the purpose of monitoring it was proposed to conduct long-term measurements of the ground temperature coupled with detailed vegetation counts according to the standard protocols at four summits located in the ecotones above the tree line.

In general, the altitudinal zonation of vegetation within the Katunskiy range is as follows:
- Within the subalpine belt sparse forests of *Pinus sibirica* and *Larix sibirica* are combined with tall-herb grasslands and subalpine dwarf-shrub heaths above the tree line ecotones.
- The alpine belt, located at altitudinal limits between 2,000 and 2,700 metres above sea level, is characterised by a dominance of alpine grasslands and high-altitudinal tundra communities, represented in the lower parts by chionophilous and semichionophilous grasslands with *Aquilegia glandulosa* and *Dracocephalum grandiflorum*, and by low-herb communities with *Sibbaldia procumbens* and *Salix turczaninowii*.
- The nival belt is fragmented and characterised by scant patchy vegetation communities with *Saxifraga oppositifolia*, *S. terektensis*, *Rhodiola coccinea* and other species combined with lichen-covered boulder fields and rocks (Artemov & Korolyuk 2001).

Biosphere reserve staff at GLORIA summit – installing temperature data loggers (© Tatjana Yashina).

The climate conditions of the alpine zone in the Katunskiy mountain range are quite severe: The mean annual temperatures observed at Kara-Tyurek meteorological station at 2,600 metres above sea level is -6.3°C, the annual precipitation is approx. 500 millimetres. Soya et al. (2006) provide an analysis of climate data from Kara-Tyurek weather station, showing a warming of approx. two degrees in winter and about half a degree in the summer season (see Tab. 1).

Time period	Tw	ΔTw	Ts	ΔTs	ΔL5	Pa	ΔPa
Before 1960s	-15.1		5.2			752	
1961-1998	-13.2	+1.9	5.7	+0.5	+6	882	+130

Tab. 1: Climate change data from Kara-Tyurek Meteorological Station (Soja et al., 2006).

(Tw –January temperarture, Ts – July temperature, Pa – annual precipitation, L5 – days with temperature continuously above +5)

The GLORIA target region (RU-AKA) is located in the western part of the Katunskiy mountain range within the core zone of the biosphere reserve. It comprises four summits which contain four different ecotones (see Tab. 2).

Summit	Altitude (m.a.s.l.)	Vegetation zone	Number of species recorded in 2005-06
Ameli	2,181	Tree line ecotone	102
Alija	2,231	Lower alpine zone	82
Prosvet	2,358	Typical alpine zone	73
Lada	2,475	Upper alpine zone	48

Tab. 2: Summits of the GLORIA target region in the Katunskiy BR.

The list of plant species/subspecies in the target region contains 138 taxa of higher vascular plants. Thirteen of these are endemics or sub-endemics of the Altai-Sayan floristic province. This province includes the mountains of Southern Siberia, Northern Mongolia and Eastern Kazakhstan. Almost all endemics recorded on the summits are widely distributed over the territory of the Altai-Sayan ecoregion. These are, for example, *Deschampsia altaica*, *Hedysarum austrosibiricum*, *Oxytropis alpina*, *Aconitum krylovii*. An exception is *Erigeron altaicus* with narrow distribution within Central Altai. The majority of the endemics can be considered neo-endemics. Their origin is related to recent changes of relief, climate and vegetation during the Pleistocene. Such species are mostly presented in the modern flora by closely related and rather similar taxa. Paleo-endemics are characterised by taxonomic isolation within the modern flora

and clearly separated from their closest relatives, among them *Saussurea frolowii* which was recorded in the target region.

Since GLORIA sites are established as long-term monitoring plots in which the vegetation surveys are repeated every ten years, it is not yet possible to determine the trends of changes in the alpine environment. However, it has been possible to determine the baseline data on species richness, composition, abundance as well as climatic conditions of alpine communities for the Katunskiy mountain range (see Tab. 3). The conditions of plant growth in the alpine environment depend on the length of the vegetation period with temperatures above +50°C, as well as the length of the snow-cover period. It would be possible to use this data as a basis for further temporal analysis of the dynamics of alpine communities.

| Summit | Altitude (m.a.s.l.) | Aspect | Number of days | | | | | | | | | Species richness |
| | | | with snow cover | | at temperature > 5°C | | | at temperature > 10°C | | | |
			2006	2007	2006	2007	2008	2006	2007	2008	
AME	2,181	E	240	225	106	97	99	90	85	83	56
AME	2,181	W	228	203	104	100	87	91	99	83	64
AME	2,181	S	201	159	130	169	122	113	97	70	65
AME	2,181	N	254	258	89	90	71	59	68	61	63
ALI	2,231	S	214	206	106	107	110	90	86	--	65
ALI	2,231	W	254	255	92	86	76	60	68	70	39
ALI	2,231	E	224	211	105	100	104	101	99	85	62
ALI	2,231	N	277	292	61	51	52	49	41	47	65
PRO	2,358	N	264	246	70	75	76	39	39	67	38
PRO	2,358	E	224	2009	92	89	89	71	77	79	50
PRO	2,358	S	236	229	103	101	102	86	83	76	49
LAD	2,475	E	285	271	60	58	--	49	52	--	24
LAD	2,475	S	219	239	101	95	97	89	83	86	30
LAD	2,475	W	233	238	108	93	97	75	79	78	24

Tab. 3: Climatic conditions of the plant growth and species richness of the alpine ecosystems of the Katunskiy BR (GLORIA RU-AKA Target Region).

Towards a GLORIA Master Site

Alpine ecosystems are not the only ones sensitive to changes in climate in high-altitude environments. The monitoring of alpine ecosystems should therefore be embedded into the wider context of other types of landscapes. Consequently, with support from the UNDP-ICI Project 'Extension of protected areas network for conservation of the Altai-Sayan Ecoregion' the complex monitoring programme for the Katunskiy BR has been developed and is being implemented. This programme includes the recording of temperatures and precipitation along altitudinal gradients as well as monitoring the upper timber line, tree line and the dynamics of glaciers. All these observations are being conducted in the GLORIA target region and will provide complex data on the dynamics of climate conditions and ecosystem response to climate change in the Katunskiy BR.

Extension of the GLORIA Network

The experience gained by the Katunskiy BR in the implementation of the GLORIA Multi-Summit Approach has facilitated the work to extend this network to other areas of the Altai-Sayan Ecoregion. Compared to other monitoring techniques, the GLORIA method has a number of advantages, such as simplicity, cost-effectiveness, standardisation of observations and recording procedures and centralised data storage. Therefore, the GLORIA approach was recommended for conducting the monitoring of high-altitude ecosystems in the protected areas of the Altai-Sayan ecoregion. The UNDP-ICI project 'Extension of the protected areas network for conservation of the Altai-Sayan Ecoregion' has supported a number of activities, including:

- The translation and publication of the GLORIA Field Manual into Russian,
- The implementation of a training-seminar for protected area staff on the GLORIA method held in 2010 in the Katunskiy Biosphere Reserve,
- The establishment of new GLORIA sites in other biosphere reserves of the Russian portion of the ecoregion, namely Altaiskiy, Sayano-Shushenskiy and Ubsunurskaya Kotlovina.

These activities will help to establish a regional network of monitoring climate change and ecosystem response based on the network of biosphere reserves.

Endemic species of the Central Altai: Erigeron altaicus (© Igor Artemov).

References

UNESCO-MAB (2008). Madrid Action Plan for Biosphere Reserves. Available at http://unesdoc.unesco.org/images/0016/001633/163301e.pdf (accessed 07-09-2010)

Galakhov, V.P. & Mukhammetov, R.M. (1999). Glaciers of the Altai [in Russian].

Kharlamova, N. (2010). Trends of climate change in the region. In: Climate Change and Connectivity Conservation in the Altai-Sayan Ecoregion. In press [in Russian].

Narozhny, Yu.K., Nikitin, S.A. et al. (2006). Glaciers of Belukha Massif (Altai): Mass balance, dynamics and distribution of ice. In: Proceedings of glaciological studies. Vol. 101, 2006, pp. 117 – 127 [In Russian].

Patrusheva, T. (2010). Dendrological indication of climate change in the high-altitudinal zone of the Altai. In: Climate Change and Connectivity Conservation in the Altai-Sayan Ecoregion. In press [in Russian].

Pauli, H., Gottfried, M. et al (eds) (2004). The Gloria Field Manual. Multi-Summit Approach. Luxembourg: Office for Official Publications of the European Communities. Available at http://www.gloria.ac.at/downloads/GLORIA_MS4_Web_english.pdf (accessed 07-09-2010)

Artemov, I.A., Korolyuk, A.Yu. et al. (2001). Flora and Vegetation of Katunskiy Reserve [in Russian].

Soja, A.J., Tchebakova, N. et al. (2006). Climate-induced boreal forest change: Predictions versus current observations. Global and Planetary Change, doi:10.1016/jgloplacha.2006.07.028.

Excursion to a mountain meadow in Großes Walsertal BR (© BR Management Gr. Walsertal).

First Steps towards Social Monitoring in Biosphere Reserves

by Sigrun Lange

From the mid 1970s, UNESCO biosphere reserves (BRs) have been established worldwide as logistic bases for the implementation of an interdisciplinary research programme on the interrelation between man and the biosphere. Thus, from the very beginning, coordinated research activities and environmental observation played a central role. In 1995, the Seville Strategy stressed that the World Network of Biosphere Reserves should be used *'as priority long-term monitoring sites for international programs, focused on topics such as terrestrial and marine observing systems, global change, biodiversity and forest health'* (Objective III.2). Back then, the majority of monitoring activities in biosphere reserves was still dedicated to nature conservation issues (biotic, abiotic). For a long time, social monitoring has been widely neglected (UNESCO 2002). However, in the course of the development from mere conservation-oriented sites to model regions for sustainable development, integrated monitoring became a key activity to be undertaken in biosphere reserves (BRIM, Biosphere Reserve Integrated Monitoring). Without social monitoring it is difficult to assess how the well-being of social groups develops, and how local people perceive the implementation of the MAB concept in their region.

Efforts to strengthen social monitoring

Monitoring refers to information or data sampling which is repeated at certain intervals and serves specific scientific and/or management purposes. It differs from pure observation or from surveys in terms of its repeatability that permits comparisons over time and evaluation against a target. Monitoring is not an end in itself but should be undertaken to achieve specific objectives. It provides scientists with socio-economic, biological or environmental data, and identifies trends. The results may assist managers or other decision-makers in implementing sustainable use and nature conservation. Nevertheless, regular data sampling in mountain biosphere reserves is carried out mainly for the assessment of biotic (e.g. species identification in GLORIA sites) or abiotic (e.g. temperature and precipitation measurements) features. In the course of two workshops in September 2001 in Rome, it was emphasised that socio-economic monitoring should complement the existing monitoring activities in biosphere reserves. In the long run, social indicators should be established and monitored regularly in order to allow for adaptive management and well-informed decision-making processes. By drafting the GLOCHAMORE Research Strategy (MRI 2005), scientists called for assessing how global change impacts human health and prosperity in mountain regions. They suggested, for example, that observations should be made regarding the distribution of organisms which cause health risks; the availability of important forest products; changes in mountain pasture conditions; or the impacts of global change on the tourism sector.

However, there are quite a lot of barriers to the implementation of sound social monitoring systems, such as insufficient funds or a lack of political will (UNESCO 2002). A new research initiative in Germany now focuses on the development of appropriate indicators and methods which allows the monitoring of social processes in BRs with comparatively little effort – at least if the monitoring is carried out in cooperation with partner universities (Stoll-Kleeman et al. 2010). The monitoring activities should allow for detecting trends in human interactions with nature and potential conflicts between sustainable use and conservation, as well as for understanding the reasons for acceptance or resistance. Four lowland biosphere reserves in Germany have been selected as pilot areas for the study (Mittelelbe, Schaalsee, Südost-Rügen, Schorfheide-Chorin).

Studies on local perception in three biosphere reserves

Experiences from the Biosphere Reserves 'Val Müstair – Parc Naziunal' in Switzerland (nominated in 2010), 'Großes Walsertal' in Austria (2000), and 'Rhön' in Germany (1991) show that studies on the perception of local people reveal the local opinion and the willingness to actively participate in the implementation of the MAB concept. So far, these studies have been carried out only once in each of the three parks. However, it is intended to repeat them over several years at irregular intervals. In the Rhön BR a second survey (using the same methodology as the first) will be implemented late in 2010; in Großes Walsertal BR this will be done as soon as the necessary funds can be acquired (probably in 2011).

The case study of Val Müstair – Parc Naziunal BR

The Müstair Valley is situated in the East of the Swiss Canton of Grisons. The Swiss National Park directly adjoins and extends to the north-west of the valley. In 1979, at the beginning of the MAB Programme, the national park was appointed the first biosphere reserve in Switzerland. However, the wilderness concept adopted for the park was always inappropriate for implementing the requirements of the Seville Strategy. Eventually, the idea emerged to combine the two areas in one biosphere reserve. In 2005 a referendum took place in the six communities of the valley. The vast majority of the local population (88.7%) voted in favour of a joint BR; only a minority of 7.8 per cent were opposed, and 3.6 per cent abstained (Corporaziun Regiunala Val Müstair & Schweizerischer Nationalpark 2005). Finally, in June 2010 the existing 'old generation' BR was considerably enlarged and re-named as 'Val Müstair – Parc Naziunal BR'. It now comprises the national park area, forming the core zone, and the community Val Müstair with approximately 1,605 inhabitants (constituted in 2009 by merging the six formerly separate communities in the valley) representing the transition zone. In line with UNESCO´s requirements, a management plan has to be developed for the entire area by 2013. In 2007, when the establishment of the BR was still in the planning phase, a survey was carried out among 191 local inhabitants in all six communities and 178 German speaking tourists in different locations (Karthäuser 2009). All interviewees had to complete standardised questionnaires in written form. This survey was complemented by 19 guided interviews with selected representatives of relevant stakeholder groups both within and outside the mountain valley. The study revealed that almost all people in the valley (95.3%) and about half of the tourists surveyed (48.9%) had already been aware of the plans to establish a BR. The most important sources of information were information events, media articles (e.g. a monthly regional newspaper which is sent to all households in the valley) and, of course, word of mouth. The visitors to the valley were predominantly informed by tourist information material and by talking to hotel staff. Whereas residents and outside experts considered UNESCO biosphere reserves to be instruments for a sustainable development of the region, the tourists mainly associated the term with nature conservation. The beautiful landscape (73.3%), recreation (49.9%) and hiking possibilities (47.7%) were amongst the main motives for visitors to come to the region. In view of the establishment of the new Val Müstair – Parc Naziunale BR, the majority of respondents (63.4% of the residents and 81.1% of the tourists) expected positive changes. By comparison, the experts surveyed wished mainly for an increase in the level of awareness and nature-based tourism leading to economic revival combined with job creation. It was a striking observation to realise that during the referendum in 2005, 88.7 per cent of the local population still voted in favour of the BR. Two years later, in the course of the survey, 'only' 63.4% of the residents expressed an expectation of positive changes from the establishment of the UNESCO site. Obviously, the time-consuming process of establishing the BR in the Swiss valley transformed some of the initial enthusiasm into impatience, doubts and a lack of understanding (Karthäuser 2009).

The case study of Großes Walsertal Biosphere Reserve

The Große Walsertal is a vast valley located in Western Austria in the federated state of Vorarlberg. The BR was designated in the year 2000; it comprises six communities with about 3,500 inhabitants. The mountain valley represents a marginalised region with low economic potential. About 800 of the economically active people commute to neighbouring regions, such as the Rhine Valley or Walgau (Weixlbaumer & Coy 2009). The most important asset is the beautiful landscape – little villages in the valley framed by mountains with altitudes of up to 2,700 metres. Despite this high potential, tourism has to compete with neighbouring destinations, such as Montafon or Bregenzerwald. Compared with those regions, the Great Walser Valley is characterised by a lower price level, lower quality standards, lack of tourism infrastructure (especially in winter), and the absence of mass tourism. From the very beginning, the establishment of the BR was seen as an appropriate tool for encouraging environmentally sensitive tourism and for developing new job opportunities for locals in the valley.

In 2005, a study was carried out by students of the University of Vienna (Austria) with the aim of analysing the perception of the BR within (self-perception) and outside the valley (external image). A total of 532 households, more than the seventh part of the total population, were targeted with a standardised questionnaire. In addition, 14 personal interviews with external experts were held. The results, presented in the table below, revealed that both the locals in the valley and the outside experts consider the designation of the biosphere reserve as having positive impacts (cf. Tab.1).

Self-perception (532 households)	External image (14 expert interviews)
• 84% consider the BR a reasonable institution	• The development of the BR was perceived as quite positive
• 68% declare having observed positive changes since the establishment of the BR	• The requirements of the Seville Strategy are fully implemented
• 40% expressed their interest in future cooperation with the BR management	• The implemented projects are considered best practice

Tab. 1: Self perception and external image of Großes Walsertal BR (results according to Weixlbaumer & Coy 2006, 2009).

In the course of the same study, the impact of the UNESCO label on tourism was analysed by students from the University of Innsbruck (Austria). Standardised questionnaires were completed by 37 guest houses which are labelled as 'partners of the biosphere reserve' and by another 46 guest houses without this label. This means that more than three quarters of the total number of guest houses in the valley have been covered by the survey. Since the establishment of the biosphere reserve, the development of half of the guest houses surveyed was positive, another third observed no changes, some recorded negative trends. In general, the guests' place of origin has not changed. Additional interviews with 169 visitors showed that the UNESCO label was not a decisive factor in their choice of destination. Half of the guests surveyed declared that they would prefer to stay in a BR-partner guest house. It must be borne in mind, that technical excursions from other biosphere reserves or by scientists – which have taken place since 2000 – impact on the visitor portfolio.

The case study of Rhön Biosphere Reserve

The low mountain ranges of the Rhön, Germany, cover three Länder (or federated states), Thuringia, Bavaria, and Hesse; and so does the Rhön Biosphere Reserve which was designated in 1991, after the unification of Germany. Each of the three parts still has its own management offices. However, a framework management plan has been established for the entire UNESCO site, with the participation of stakeholders from all three Länder. In 2002, a study was commissioned in order to record the general opinion of the public with regard to the BR and the activities of the management (IfD 2002). A total of 61 per cent of the respondents associate the Rhön BR with nature conservation (species or landscape conservation). The BR enjoys a good reputation with three quarters of respondents; 72 per cent assume that the BR yields benefits for the region (e.g. for raising the degree of popularity of tourism). About two thirds (62%) believe that the biosphere reserve strengthens the regional identity across the boundaries of the three Länder. While 71 per cent know the visitor centres in the BR, almost one third (29%) have already visited one of them and were quite pleased with the information they received. However, more than half of the interviewees (58%) had the impression that the decision-making process on behalf of the management is not always transparent and about one third (31%) complained about too many restrictions (IfD 2002). In addition to monitoring the general opinion of the public, in 2008, the Rhön BR released its first integrated environmental report (BayStMUGV, HMULV & TMLNU 2008), reporting on the period 1991–2006. Even if the state of the environment is the main focus of this report, socio-economic data has been included, comprising for example

- the demographic development (e.g. population decline in Thuringia, increase in the greater conurbation of the Hessian city of Fulda);
- the job market (e.g. decline in jobs subject to social insurance contributions in most of the Rhön counties, with the exception of the city of Fulda);
- agriculture (e.g. high percentage of organic farming);
- tourism (e.g. decline in bed nights);
- the settlement development (e.g. increase in residential building areas despite declining population figures);
- traffic (e.g. traffic volume below the regional average); and
- energy (e.g. percentage of renewable energy above the regional average).

This survey is to be linked to the 10-yearly evaluation period and report on the development of the BR and its achievements with respect to environmental protection and curbing the key drivers of its deterioration.

References

BayStMUGV, HMULV & TMLNU (eds.) (2008). Erster integrierter Umweltbericht für das länderübergreifende UNESCO-Biosphärenreservat Rhön.

IfD – Institut für Demoskopie Allensbach (2002). Biosphärenreservat Rhön – Allensbacher Repräsentativbefragung im Frühjahr 2002.

Karthäuser, J.M. (2009). Die Biosfera Val Müstair – Parc Naziunal: Zur Akzeptanz des geplanten UNESCO-Biosphärenreservats. In: Mose (Hrsg.) (2009). Wahrnehmung und Akzeptanz von Großschutzgebieten. Wahrnehmungsgeographische Studien, Band 25: 83–108.

MRI (ed.) (2005). GLOCHAMORE – Global Change and Mountain Regions Research Strategy. A joint project of the Mountain Research Initiative (MRI), UNESCO-MAB and IHP, and the 6th EU Framework Programme.

Stoll-Kleemann, S., Buer, C., Hirschnitz-Garbers, M. & Solbrig, F. (2010). Instrumente für ein künftiges soziales Monitoring in deutschen UNESCO-Biosphärenreservaten. Bericht des Workshops am Lehrstuhl für Nachhaltigkeitswissenschaften und angewandte Geographie der Universität Greifswald am 6. & 7. Mai 2010.

UNESCO (2002). Social Monitoring – Meaning and Methods for an Integrated Management in Biosphere Reserves. Report of an International Workshop, Rome, 2–3 September 2001. Biosphere Reserve Integrated Monitoring (BRIM) Series No 1. Paris, France.

Weixelbaumer, N. & Coy, M. (2009). Selbst- und Fremdbild in der Gebietsschutzpolitik. Das Beispiel des Biosphärenparks Großes Walsertal/Vorarlberg. In: Mose (Hrsg.) (2009). Wahrnehmung und Akzeptanz von Großschutzgebieten. Wahrnehmungsgeographische Studien, Band 25: 37–57.

Weixelbaumer, N. & Coy, M. (Hrsg.) (2006). Zukünftige Entwicklungsstrategien für den Biosphärenpark Großes Walsertal. Eine regionalwirtschaftliche und perzeptionsgeographische Analyse. Projektendbericht. Finanzierung durch die ÖAW.

Gossenköllesee BR, Austria, has a long research tradition (© Planet Austria / Lammerhuber).

A New Label for Biosphere Reserves with a Long Research Tradition? The Case of the Gossenköllesee BR, Austria

by Günter Köck

For more than ten years the International Coordinating Council of UNESCO's Man and the Biosphere Programme (MAB-ICC) has been discussing the fact that many biosphere reserves within the World Network of Biosphere Reserves (WNBR) do not conform to the requirements of the modern biosphere reserve concept as proposed in the Statutory Framework of the WNBR (UNESCO 1996). Since 2008 this discussion has become even more intense when during the Third World Congress for Biosphere Reserves the Madrid Action Plan (UNESCO 2008) was adopted. The Action Plan notes that nearly all BRs nominated since 1995 conform to the modern zonation criteria mentioned in Article 4 of the Statutory Framework of the WNBR (UNESCO 1995). However, a considerable number of the sites nominated between 1976 and 1995 are lacking the required three zones (Price et al. 2010, G. Köck, pers. comm.). This is also true for four Austrian biosphere reserves.

The history

In 1976 the world-wide network of biosphere reserves was founded. At that time, early on in the MAB programme, classical conservation thinking still prevailed. Representative ecosystems world-wide were to be protected and maintained as trial areas for internationally co-ordinated research projects. With the Seville Conference in 1995, the MAB programme underwent a significant change: The former research programme was transformed into a modern instrument for the conservation and sustainable development of regions. In a holistic concept, humans and their economic activities are integrated in the conservation of biodiversity. Research was still important but was only one aspect of the holistic concept (Lange 2005, Köck & Lange 2007).

The six Austrian biosphere reserves Neusiedler See (Burgenland), Lower Lobau (Vienna), Gossenköllesee (Tyrol), Gurgler Kamm (Tyrol), Großes Walsertal (Vorarlberg) and Vienna Woods (Vienna, Lower Austria) cover a total area of 1,518 square kilometres or roughly 1.8 per cent of the Austrian territory (Lange 2005). Four of the Austrian biosphere reserves were designated as far back as 1977: Gurgler Kamm, Gossenköllesee,

Neusiedler See and Untere Lobau. At that point, the initiative for the selection of the areas came from scientists. For many years, therefore, it was mainly basic research that went on in the new protected areas. In these 'first-generation biosphere reserves', UNESCO's international guidelines have so far been implemented insufficiently. The first 'modern' Austrian BR was created in the Große Walsertal in 2000, followed by Vienna Woods in 2005 as a further 'model region for sustainable development', a term which can certainly not be attributed to the four Austrian pre-Seville sites (Köck et al. 2009).

The Austrian MAB National Committee (MAB-NC) has recognised this critical situation and implemented new 'National Criteria for BRs in Austria' in 2006, thereby allowing a five-year transition period for the currently existing 'first-generation' sites (Austrian MAB-NC 2006). If the areas, at the end of that period, do not adequately meet the criteria, they will be withdrawn from the WNBR list. At the same time, the National Committee started 're-design' initiatives (e.g. in the BRs Neusiedler See and Gurgler Kamm) to transform its 'first-generation sites' into modern Seville-style biosphere reserves. However, this track turned out to be difficult and interminable. Let me explain the situation by describing the status of the Gossenköllesee BR, the world´s smallest biosphere reserve.

The Gossenköllesee BR problem – symptomatic of many other sites in the World Network of Biosphere Reserves

When the MAB Science Programme was initiated in the early 1970s, the 'International Biological Programme (IBP)' was about to be wound up. Until 1974, in the course of the Austrian part of the IBP, many research projects were carried out successfully in the Neusiedler See and its reed belt, and in the Tyrolean Alps. The MAB Programme in Austria was meant to continue the IBP research in extended form. Early on Austria participated mainly in the core research themes of limnology and mountain ecology under the new UNESCO MAB Programme. The initiative for setting up the four biosphere reserves therefore came mainly from the researchers. The establishment of the BR Gossenköllesee also goes back to an initiative by researchers (led by Prof Walter Moser, the Director of the Alpine Research Station Obergurgl) in 1977 who tried to secure the continued existence of an internationally relevant research site.

The Gossenköllesee, a high-mountain lake situated at 2,417 metres above sea level in the Stubaier Alps, has a surface area of 1.6 hectares. Only ten per cent of the catchment area is covered with thin soil, home to a sparse vegetation of lichen and typical plants of Alpine grassland and ericaceous dwarf shrubs. The lake is usually covered with ice from the beginning of November until the end of June. The biosphere reserve consists only of the lake and its catchment area and is, with a size of only 85 hectares, the world´s smallest biosphere reserve. The only form of land use in the area is grazing by sheep. No humans live in the catchment area of the Gossenköllesee. The nearest village in the valley is Kühtai, one of the best-known ski resorts in the Tyrol. Situated at an altitude of 2,020 metres, Kühtai is not only the highest winter sports village in Austria; with its 13 inhabitants it is probably also the least populated one in the country. A ski-lift stops at the edge of the BR.

To date, the Gossenköllesee BR is reserved exclusively for research. Since 1997 research activities on Gossenköllesee have focused on studying the effects of global ecological change on catchment areas for high-alpine waters (Psenner 2009). The well-equipped Limnological Research Station of the University of Innsbruck, which has collected climate data for more than 30 years, turned the biosphere reserve into an important centre of high-mountain research in Europe. Since 1992 the Gossen-köllesee has been part of various EU projects (ALPE, EMERGE). For instance, the lake played a central role within the international research project 'MOLAR' (Mountain Lake Research, 1997– 1999) which compared 13 European high mountain lakes. It was also integrated into the EUROLIMPACS project as part of the 'Network of Excellence' ALTER-NET within the EU's 6th Framework Programme. The Gossenköllesee is the only high-mountain lake in Europe with a well-endowed research station where equipment-intensive measurements can be taken. In 1994 the Station was modernised to ensure emission-free operation. This special infrastructure made the Gossenköllesee the place of choice for participation in a research cooperation project run jointly by the UNESCO MAB Programme and the Mountain Research Initiative (Switzerland) (GLOCHAMORE). It envisages setting up monitoring stations in mountain regions all over the world to serve as early warning systems for the effects of global climatic change or change in pollutant capture. Furthermore, the biosphere reserve is involved in numerous active international partnerships such as partnerships with universities in the United States (Montana), Spain (Barcelona), research institutions in Germany (MPI Bremen, MPI Marburg), the Czech Republic (Academy of Science, Budweis), University College London and many more (Lange 2005).

The development of an idea

It is obvious that, owing to its long history of research activities, the Gossenköllesee is a particularly valuable research site. However, the area does not fulfil the criteria of a modern biosphere reserve. The site – with 85 hectares the smallest biosphere reserve in the world – only covers the lake and its catchment area. There is a lack of comprehensive zoning. There are no people living within or close to the boundaries of the biosphere reserve. Furthermore, there is no BR management and also a lack of official government funding.

For geographical and political reasons the overall chances of transforming the area into a modern biosphere reserve are close to zero. Consequently, the only option for the Austrian MAB-NC at present is to withdraw the Gossenköllesee BR from the WNBR list. However, a withdrawal of the BR designation would not be a satisfactory option: This valuable research site provides important long-term data for environmental monitoring in alpine areas, and it is in close vicinity to a skiing area. For 20 years it has been threatened by the extension of this ski resort. If the extension plan goes ahead, new lift tracks would cross the catchment area of the Gossenköllesee. In the past, the UNESCO label has helped to protect the area from being included in the skiing area. If the area loses its designation it will most likely be covered immediately in ski lifts and tracks and will thus be lost to the research community.

In discussion with colleagues from other national committees, the Austrian MAB Committee has realised that this problem is not unique to the Gossenköllesee but is evident in many countries. Soon, it became clear that in many cases this transformation process would, for political reasons or other external constraints (e.g. limited space of area worth protecting), require very difficult and protracted negotiations about zoning, enlargement, participation and other criteria, or would be simply impossible. However, despite their old-fashioned framework

Climate station at Gossenköllesee (© Planet Austria / Lammerhuber).

many biosphere reserves may have a high societal value and/or have a very long tradition as scientific research site where excellent long-term data series are available on a wide range of scientific topics. A withdrawal of the designation would be counterproductive in this case, especially as the Madrid Action Plan calls for using UNESCO's WNBR for monitoring the effects of global change. To overcome this unsatisfactory situation, the Austrian MAB-NC started a discussion process. This gave rise to an idea which would benefit both the biosphere reserve and the MAB Programme.

A new label under discussion

It is safe to assume that a fairly high number of 'first-generation' style BRs whose future status is still unclear exist in many countries. The Austrian MAB Committee therefore suggested the definition of a new category of protected areas within the MAB Programme for 'first-generation' BRs that cannot be converted into modern biosphere reserves but can provide evidence for exceptional social or scientific value as 'MAB scientific research sites' or 'MAB conservation sites' (name to be discussed). The proposal was first announced during the MAB-ICC Meeting in 2006 and again two years later at the MAB-ICC Meeting in Madrid and led to initial discussions of this option. Already at the 21st Session of the MAB-ICC in Jeju in 2009 the proposal was seconded by several countries. In 2009 I undertook a survey among members of the EuroMAB Group to ascertain the possible number of 'first-generation biosphere reserves' which cannot be transformed into modern biosphere reserves. The result (an estimated number of at least 18 BRs with an unclear future) encouraged us to proceed. Consequently, the idea was proposed again at the MAB-ICC Meeting in 2010. The proposal, supported by many other countries in oral statements, is now reflected in the official protocol of the meeting. The paragraph reads as follows:

'*52. Several ICC Member Delegates expressed their concerns regarding pre-Seville sites that cannot be transformed into the post-Seville biosphere reserve model but would still retain international significance for research and demonstration studies on issues and problems of the environment. The ICC was sympathetic to the proposal from the Member Delegate of Austria made at the 21st session of the Council in May 2009 for creating a set of "MAB research sites" separate from the WNBR but emphasized that criteria and quality standards for the inclusion of sites in a new set of MAB research sites need to be established*' (UNESCO 2010).

It was made quite clear in the discussion that a new label such as MAB research site would not mean that all first-generation research sites be abandoned, regardless of their value and importance for the MAB Programme. It is emphasised that only those first-generation sites would be eligible for the proposed new label, which can provide evidence for a long tradition of use as research sites and/or social value, thus confirming their significant value for the MAB Programme. For example, these sites would be extremely helpful in supporting the MAB Programme´s efforts regarding the implementation of the UNESCO Strategy for action on climate change. Whilst we agreed that after 2013 the World Network of Biosphere Reserves should only consist of second-generation sites, we believe that valuable pre-Seville BRs all over the world should be kept within the MAB Programme under the new category of 'MAB research sites'. In order to bring the process forward, the Austrian MAB-NC has offered to host an expert meeting in Vienna in order to discuss and elaborate a strategy for creating a set of 'MAB research sites' including non-transformable pre-Seville BRs as well as any other new sites dedicated to research in the field of global environmental issues.

The research in the Gossenköllesee BR (cp. image above, © Planet Austria / Lammerhuber) is not a typical project of two or three years' duration, but a long-term research programme committed to studying high-altitude ecosystems under the auspices of the Austrian Academy of Sciences since 1977. In the course of numerous dissertations and projects also sponsored by the Austrian Science Fund and the EU, researchers and students contributed to the understanding of problems most of which had a global background such as acid rain, deposition of pollutants, changes in biodiversity and the consequences of climate change, but also shared the characteristics of extreme habitats and communities. Alpine or high-altitude lakes are especially suitable for the study of global change, because they have no direct human impacts in their catchments and relatively simple food webs, and react very sensitively to changes in the atmosphere and the watershed. Alpine lakes, however, are not only sentinels but also archives of environmental changes which are reflected in their sediments accumulated since the early Holocene (Psenner 2009, Köck 2010).

References

Austrian MAB-NC (2006). National Criteria for Biosphere Reserves in Austria. In: Erhalt der biologischen und kulturellen Vielfalt. Modelle für nachhaltige Entwicklungsstrategien im 21. Jahrhundert. Orte der Forschung, Bildung und Umweltbeobachtung. Austrian MAB National Committee, Austrian Academy of Sciences, Vienna. Download: http://epub.oeaw.ac.at/0xc1aa500d_0x0011e796

Lange, S. (2005). Inspired by diversity – UNESCO´s biosphere reserve as model regions for sustainable interaction between human and nature. Austrian Academy of Sciences Press, Vienna. Download: http://epub.oeaw.ac.at/3596-3inhalt

Köck, G., S. Lange (2007). UNESCO Biosphärenparks in Österreich – Modellregionen für nachhaltige Entwicklung. In: PERSPEKTIVEN 7: 14–18.

Köck, G., Koch, G. & Diry, C. (2009). The UNESCO Biosphere Reserve 'Biosphärenpark Wienerwald' (Vienna Woods) – a Long History of Conservation. Eco.mont 1(1): 51–56.

Köck, G. (2010). Mountain and global change research programmes in Austria. In: Institute of Mountain Research: Man and Environment. Challenges for Mountain Regions – Tackling Complexity. Borsdorf, Grabherr & Stötter (eds.), Böhlau Verlag, Wien.

Price, M.F., Park, J.J. & Boumrane, M. (2010). Reporting progress on internationally designated sites: The periodic review of biosphere reserves. Environmental Science & Policy 13: 549–557.

Psenner, R. (2009). Long-term Research in the Biosphere Reserve Gossenköllesee. In: Planet Austria – Stein, Wasser, Leben (2009). G. Köck, L. Lammerhuber, W. Piller (eds.), Edition Lammerhuber and Austrian Academy of Sciences Press. Download: http://www.planet-austria.at

UNESCO (1996). The Seville Strategy for Biosphere Reserves and The Statutory Framework of the World Network of Biosphere Reserves. UNESCO, Paris.

UNESCO (2008). Madrid Action Plan for Biosphere Reserves (2008–2013). UNESCO, Paris.

UNESCO (2010). Final report of the 22nd Session of the International Coordinating Council of the man and the Biosphere (MAB) Programme. Download: http://www.unesco.org/mab/doc/icc/2010/e_finalRep_may.pdf

Giant beeches in the Uholka primeval forest in Carpathian Biosphere Reserve (© Myroslav Obladanyuk).

A Piece of Wilderness: The Conservation of the Primeval Beech Forests in the Carpathian Biosphere Reserve, Ukraine

by Fedir Hamor, Vasyl Pokynchereda, Victoria Gubko, Yaroslav Dovhanych

Extending over 1,500 kilometres and seven countries, the Carpathians are Europe's largest mountain range and a natural treasure of global significance. With a total length of approx. 280 kilometres, the Ukrainian Carpathians cover eleven per cent (3,700 km²) of this mountain range. They are significantly lower than other parts of the Carpathians; their summits do not reach the glacier line. With a dense network of rivers and a great number of lakes, the Ukrainian Carpathians serve as an important source for freshwater. They are also a powerful climate-forming and water-regulating factor for continental Europe.

Three ethnic groups inhabit the Ukrainian Carpathians, the Hutsuls, the Boiky and the Lemky They depend mainly on the region's natural resources, such as timber, pastures and wild fruit including berries. All of them have retained an authentic culture, but the Hutsul are known in particular for their unique wooden architecture (churchesand grazhda houses, i.e. traditional wooden houses), trades and handicrafts, authentic folklore, lively melodies and dances. Alpine sheep farming is very popular in this region. The meadows (locally called 'polonynas') are mostly owned by communities. Shepherds usually stay in the mountains for three or four months in the year. The basic product of this type of farming is cheese which is produced by shepherds using an old technology. The end of the polonynas season is celebrated as a special festival. For farmers, it is the most important event of the year.

Besides cultural diversity, the Ukrainian Carpathians still harbour many areas of near-natural ecosystems, amongst them Europe's largest contiguous beech forest, the primeval forests in the Ukrainian Uholka-Shyrokiy Luh Massif. Since 1968, these exceptional ancient forests have been protected rigorously as nature reserve ('zapovednyk'). In 2007, they were added to the UNESCO World Natural Heritage list as part of the trans-boundary Ukrainian-Slovak serial site known as 'Primeval Beech Forests of the Carpathians'. The forests and other areas of exceptional conservation value are included in the wider area of the Carpathian Biosphere Reserve (BR), which was established by a Decree issued by the President of Ukraine in 1993, based on the Carpathian Nature Reserve, which had been founded back in 1968. Furthermore, the reserve was three times (in 1997, 2002 and 2007) awarded the European Diploma by the Council of Europe, for its great contribution to nature conservation and the protection of natural and cultural heritage.

The Carpathian Biosphere Reserve

The Carpathian BR comprises the Rakhiv, Tyachiv, Khust and Vynohradiv administrative districts of the Transcarpathian region. No less than 17 settlements are located within the BR and approx. 100,000 people live within its zone of activity. Nearly 400 people live directly in the territory of the BR, mainly engaged in animal husbandry and crop cultivation (UNESCO MAB 2007). The BR's total surface comprises 58,036 hectares. It consists of eight isolated massifs that lie within the altitudes of 180 to 2,061 metres above sea level (m.a.s.l.). Owing to this territorial structure, the reserve represents practically all topological and biological diversity of the southern slopes in the Ukrainian Carpathians. Oak forests in the foothills with slight or no disturbance, as well as mountain forests of beech or spruce as well as mixed stands, subalpine and alpine meadows with stunted pine-alder stands and rock-lichen landscapes are protected here. The dominant vegetation type is forest, which occupies 82 per cent of the total area and covers the territory starting from the foothill zone (250 m.a.s.l.) up to the alpine forest belt (1,720 m.a.s.l.). Approximately 33,000 hectares of the forests in the Carpathian BR are described as natural; of which 20,000 hectares are considered to be primeval forests. With an expanse of 24,736 hectares (56.1%), broad-leaved forests slightly outweigh the coniferous ones (19,371 ha, 43.9%). Pure and mixed beech (22,593 ha) and spruce (17,813 ha) stands dominate. The BR's forest ecosystems impress with an over-whelming diversity: No less than 33 forest types and 245 plant and animal communities are described here, 55 of which are listed in the Green Book of the Ukraine (Didukh 2009). A characteristic peculiarity of the BR's forests is the great abundance of old hollow trees which shelter a number of animal species.

The BR hosts a high number of plant and animal species, many of which are rare and endangered. In general 3,029 plant species (amongst them 1,359 species of higher vascular plants), 308 vertebrates (66 of which are mammals). No less than 193 bird species, nine reptile species, 14 amphibian species, 27 fishes, one cyclostome and approx. 15,000 invertebrates have been identified within the BR. Of these, 181 plant and 132 animal species are listed in the Ukraine's Red Book (Akimov 2009). Large carnivores, such as lynx (*Lynx lynx*), wolf (*Canis lupus*) and bear (*Ursus arctos*) are still present in this mountain region. A permanent but very rare inhabitant is wild cat (*Felis silvestris*). European mink (*Mustela lutreola*) is found in a few places, although it has practically disappeared from its natural range. Here, this animal dwells on the banks of small mountain streams. The otter (*Lutra lutra*) is also often encountered in such locations; it is the mink's greatest competitor. Both species are listed in the Ukraine's Red Book of (Akimov 2009).

From strict protection towards sustainable development

Since the designation of the BR, the functions of protected areas and their role in relation to local stakeholders have changed significantly. Nowadays, the Carpathian BR is managed according to the principles outlined in the Framework Convention on the 'Protection and Sustainable Development of the Carpathians', adopted in 2003 at the Ukraine's initiative. The BR administration is well aware that the strategy for nature conservation in the Carpathians has to be changed. In Soviet times, a reserve ('zapovednik') meant strict protection; any kind of human intervention was restricted. Normally, a zapovednik was smaller in size (as it was not possible to withdraw large areas from human use), which did not allow the conservation of contiguous ecosystems. Of course, it was not intended to take the local people's interests into account. Since the Carpathian Nature Reserve became a BR with the associated objective of harmonising the relationship between humans and nature, the zoning has changed. Today there are four zones:

- the core zone (31%), assigned to strict conservation and scientific research; only few areas are open to visitors in order to avoid pressure on the most precious ecosystems which constitute the heart of the BR;
- the buffer zone (28%), where some regulated activities are permitted;
- the zone for traditional forms of utilising natural resources (35%); and
- the regulated conservation zone (6%), established along hiking trails entering the core area with associated regulations for visitors.

A natural treasure – the primeval beech forests

The Uholka-Shyrokyi Luh Massif is the world's largest massif covered in virgin beech forest (*Fagus sylvatica*). Its total area comprises 15,033 hectares, of which 8,835 hectares are consigned to the core zone of the BR and 6,198 hectares to the buffer zone. The local soil (limestone) and climate conditions fulfil the ecological requirements of *Fagus sylvatica*; the beech forests have achieved their highest successional state – the climax stage. Starting from 380 metres up to the upper tree line (ranging from 1,250 to 1,350 m.a.s.l.), beech forest makes up a continuous vegetation zone. 360 year-old trees are rather common here. Some of them reach 55 metres in height with a diameter of approx. 1.4 metres. Giant beech trees whose thick grey candle-like trunks are deprived of branches up to a level of 30–40 metres, form a giant colonnade in this majestic temple of nature. In the core zone, these beech forests are preserved in their pristine state. A great number of standing and fallen deadwood covered with carpets of moss and beards of lichen make up an inimitable 'harmonious chaos' which allows our contemporaries to see the world with Herodotus's eyes. No one is left unmoved by the spectacle of these primeval forests. For decades visitors from all over the world have been deeply impressed and bewitched by their prehistoric beauty. In the beech forests of the buffer zone it is permitted to collect mushrooms and berries, to have tourism and recreation activities, and to establish campsites. In the buffer zone logging is restricted; only forest hygiene operations and nature conservation measures are permitted.

Hunting regulations within the Carpathian BR

An important objective for biosphere reserves is the support of traditional forms of land use. However, one of the traditional forms of utilising natural resources – hunting – is currently NOT permitted. Ukrainian legislation prohibits this kind of activity within protected areas of any rank. Legitimate hunting is therefore only practised within state-forestry enterprises. In most European countries, the population density of ungulates (*Cervus elaphus*, *Capreolus capreolus*, *Sus scrofa* etc.) is much higher than in the Ukrainian Carpathians. In Austria, for example, this number reaches 100 individuals per 1,000 hectares of a hunting site (Dyozhkin 1983). This is explained as being due to the hunters' acute interest in safeguarding an abundance of wildlife. Furthermore, there are practically no large carnivores as natural regulators of wild ungulates. However, high populations of game causes problems to forestry, because animals greatly damage forest plantations, in particular

Autumn colours in the forests of Carpathian BR (© Myroslav Obladanyuk).

young stands. In the Carpathian BR, however, wildlife population densities are much lower than the optimum. In 2009, the population density per 1,000 hectares was 4 to 5 individuals for red deer, 5 to 6 for roe deer, and 4 to 5 individuals for wild boar (Hamor et al. 2009). According to Turianin (1975), this differs significantly from the optimum for the Ukrainian Carpathians as a whole, ranging from 10 to 25 red deer, 15 to 50 roe deer and 10 to 20 wild boar per 1,000 hectares, depending on the quality of a given site. Inventory data prove that there is practically no difference between the population densities of roe deer and red deer, both within and outside the BR. This is explained by the habit of these species of using the food base evenly. As for wild boar, its population density is somewhat higher in the BR (especially in primeval beech forests) than in the adjacent areas. The low game population density is primarily due to poorly organised hunting and secondly a high level of poaching. Moreover, some parts of the ungulate populations have been depleted by carnivores, in particular wolves which are still roaming the Ukrainian Carpathians. The management of large carnivores (e.g. by hunting) and some ungulates might become necessary in the Carpathian BR, but only if their number exceeds the scientifically justified population limits. However, this can only happen in case populations expand throughout the Ukrainian part of the Carpathian Mountains. To this end, game conservation would have to be strengthened and regulations for legitimate hunting would need to be improved.

Research and monitoring

The Carpathian BR is one of the greatest scientific and educational centres in the Carpathian Region. There are five scientific laboratories, a network of monitoring plots, phenological stations, hydrological and weather stations, as well as a GIS laboratory. The BR serves a number of national and international research institutions as a natural laboratory. The Carpathian BR actively cooperates at international level. The institution has a ten-year history of cooperating with the Swiss Federal Institute for Forest, Snow and Landscape Research (WSL). Jointly with this important Swiss research institution, the management team has implemented a number of interesting activities linked with forest research; above all, the exploration of primeval forests. The project of establishing a large-scale inventory of the primeval beech forests is now well under way[1]. The Carpathian BR also works in close cooperation with the Eberswalde University for Sustainable Development, Germany. The project on improving the management strategy for the BR is also under way. The basic task of the project is to optimise the practical management of the BR by using a conceptual model which is developed on the basis of stakeholders' visions. It will stimulate a more participatory and proactive management approach. To facilitate this process, it was proposed to use the 'Miradi Software'[2] for the adaptive management of conservation projects. Now the Ukrainian version of the software is being produced internally within the project. It is planned to distribute the outcome among other protected areas in the Ukrainian Carpathians and in the Ukraine in general. Another project was carried out regarding the identification of primeval forests for the whole Transcarpathian region together with the Royal Dutch Society for Nature Conservation (Hamor et al. 2008).

There is an intensive exchange of students between prominent scientific institutions of the Ukraine and foreign countries (e.g. Eberswalde University for Sustainable Development, Germany, and others). The Carpathian BR hosts international conferences almost annually, such as
- 'The Carpathian region and problems of sustainable development' (October 13-15, 1998),
- 'Mountains and people (in the context of sustainable development)' dedicated to the International Year of Mountains (October 14-18, 2002),
- 'Natural forests in the temperate zone of Europe: values and use' (October 12-17, 2003),
- 'Ecotourism and sustainable development in the Carpathians' (October 9-12, 2007), and
- 'Protected area system development in the Ukraine and formation of the pan-European ecological network' (November 11-13, 2008).

In addition, there is a whole network of educational and scientific information trails and information centres in the BR: the Museum of Mountain Ecology and the Narcissus Museum, a photo and video studio. The Ukrainian-wide journal 'Zeleni Karpaty' (Green Carpathians) and the Carpathian BR Newsletter are published by the BR.

Transboundary cooperation

The Maramures Massif, protected within the Carpathian BR, directly adjoins the Romanian Maramures Mountains National Nature Park. Its geographical location and the active cooperation between the two protected areas make up a fundamental platform for the future establishment of a transboundary BR in the Maramures Mountains based on the BR and adjacent National Park on the other side of the border. This idea arose long ago in the Administration of the Carpathian BR, and now we are at the stage of preparing the joint nomination to the UNESCO MAB Secretariat to apply for a Transboundary Biosphere Reserve. It is obvious that nature knows no boundaries. A shared ecosystem therefore demands joint conservation and management measures despite being divided by the EU border.

Facing the challenges of the future

The region where the BR is located, is facing rapid socio-economic development and has been undergoing many changes and transformations since 1991. Factors impacting on the region include the de-collectivisation of agriculture and forestry, high unemployment rates and work-related migration, land privatisation, inflation and global developments such as climate change (Geyer, Hamor, Ibisch 2009). Unsustainable forest use and

---

[1] Information on the WSL inventory project at http://www.wsl.ch/fe/walddynamik/projekte/uholka/index_DE

[2] Adaptive Management Software for Conservation Projects: https://miradi.org/

illegal logging results in continued loss of older forests and their services as well as in the ongoing fragmentation of some of Europe's last large mountain forests as found in and around the Carpathian BR (Kuemmerle et al. 2009).

The galvanisation of local interest groups into a unit for decision-making regarding the management of protected areas has been rather deficient in the past (just like in other post-socialist European countries) and the high dependence of various players on natural resources for their business and livelihood gives rise to conflicting interests and low acceptance of the BR among the population (Wallner 2005). As management practices are more likely to be accepted and implemented when key players have been involved in the decision-making process, the Carpathian BR recently started to actively involve the local public. This cooperation is achieved mainly through the 'Coordination Board' which meets annually or if there is an urgent issue to be discussed. It consists of management staff members, representatives from local communities, local authorities and other important stakeholders. The board discuss issues important both to the protected area and the communities, and try to find solutions for conflicts where they cannot be avoided. For example, local communities are allowed to collect firewood in anthropogenic landscapes, and to use pastures in the buffer zone and the zones where the traditional utilisation of nature is permitted. The Carpathian BR sells firewood at lower prices if it is the product of harvesting in the course of sanitary logging and nature conservation measures. The BR also provides a possibility for the traditional utilisation of nature such as harvesting of non-timber forest products where it is legally permitted.

As mentioned before, local people of the area make a living mainly from natural resources. It is obvious, therefore, that any restrictions in connection with nature-conservation regimes will cause conflicts of interests. That is why representatives from the Carpathian BR Administration try to find a common language with the stakeholders by attending village and town council meetings, paying visits to state forestry enterprises and establishing both official and personal contacts with important players. The principles of zoning, restrictions and opportunities are explained to the public through the organs of mass media both at local and regional level. Stakeholder involvement is a real challenge, but the Carpathian BR is effective in communicating its main objectives to people and in explaining new opportunities to them. As a result, it was possible to celebrate another achievement in 2009: the BR was extended by another 7,000 hectares by Presidential decree, but of course, this is the outcome of an extended process of discussions and agreements with local authorities and communities.

References

Akimov, A.I. (ed.) (2009). Redbook of Ukraine.

Didukh, Ya. (ed.) (2009). Greenbook of Ukraine, Natioual Academy of Sciences of Ukraine.

Dyozhkin, V.V. (1983). Hunting management in the world.

Geyer, J., Hamor, F.D. & P.L. Ibisch (2009). Carpathian Biosphere Reserve (Ukraine): Towards Participatory Management. In: eco.mont – Volume 1, Number 2, December 2009.

Hamor, F., Pokynchereda, V. et al. (2009). Chronicles of Nature of the Carpathian Biosphere Reserve, Volume 2009.

Hamor, F., Pokynchereda, V., Dovhanych, Ya. et al. (2008). Primeval forests of Transcarpathia – Inventory and Management.

Kuemmerle, T., O. Chaskovskyy, J. Knorn, V.C. Radeloff, I. Kruhlov, W.S. Keeton & P. Hostert (2009). Forest cover change and illegal logging in the Ukrainian Carpathians in the transition period from 1988 to 2007. Remote Sensing of Environment 113, 6: 1194–1207.

Turianin, I. (1975). Game fauna and animals important for fir production in Transcarpathia.

UNESCO MAB (2007). UNESCO's database on the world net of biosphere reserves at: http://www.unesco.org/mabdb/br/brdir/directory/biores.asp?mode=all&code=UKR+03 (accessed on 24 October 2010).

Wallner, A. (2005). Biosphärenreservate aus der Sicht der Lokalbevölkerung. Schweiz und Ukraine im Vergleich. Dissertation. Birmensdorf.

Worldwide Case Studies

The Challenge of Reconciling Conservation and Sustainable Development

Learning from experiences

Chapter 3-2

Sacred Dad Alchi's Cedar in Altaiskiy Biosphere Reserve (© E. Veselovskiy).

Altaiskiy Biosphere Reserve: Indigenous and Local People's Contribution to Conservation and Sustainable Development

by Yuri Badenkov, Svetlana Shigreva & Igor Kalmikov

Altaiskiy Biosphere Reserve

The Altaiskiy Biosphere Reserve (BR) is located in the area of the north-eastern and eastern Altai and occupies the eastern part of the Teletskoye Lake basin (Fig 1). Plateaus and alpine ridges of the Chulyshmansky highland occupy the greater part of its territory. The average altitude of the mountains is approx. 1,900 metres; the highest summit reaches 3,148 metres. The deep valley of the river Chulyshman and the Teletskoye Lake (at 434 m.a.s.l.) extend along the western border of the BR. The mountainous area is the origin of several major rivers. For example, in the western section, the rivers Ob and Yenisei flow to the Arctic Ocean, and several rivers flow towards Inner Mongolia. A characteristic feature of the landscapes is the abundance of lakes. Some 2,560 medium-sized and small lakes and approx. 1,200 lakes of glacial origin are located in the BR. Swampy areas are found in inter-mountain depressions, river valleys and on smooth slopes.

The BR covers a total area of some 3,532,234 hectares. It comprises three functional zones:
- Core zone (Altaisky State Nature Reserve, covering 881,236 ha corresponding to 25 % of the total area) with the aim of conserving landscapes, ecosystems, species and genetic diversity;
- Buffer zone (962,800 ha corresponding to 27 % of the total area) with the aim of restricting any procedures or activities which have a negative impact on the core area;
- Transition zone (1,688,198 ha corresponding to 48 % of the total area) which includes agricultural, natural, semi-natural and municipal land, as well as villages of the Turochaksky and Ulagansky districts. The transition zone should allow for the region's development, in a manner that is environmentally and socio-culturally sustainable.

Indigenous people – ancient inhabitants of the area

At the end of the 19th century, the area was sparsely populated. Only a few villages of the indigenous Teles people, and several hunters' winter huts were located in the area of the reserve.

Since times immemorial, indigenous tribes such as the Teles, Tubulars, Telenghits and Kumandins, have been using the area for hunting, animal-raising and gathering cedar nuts. They are the custodians of traditions, customs and ceremonies of a culture which is closely related to a nomadic and semi-nomadic lifestyle. The Teles and Telenghit people were raising animals in the southern areas. Cattle and horses were pastured all year round. These tribes were always showing a special relationship with nature. The forests were not cleared, and only a small amount was cut for firewood; minerals were of no interest to them. The Tubulars inhabited the northern forested areas. They practised hunting and fishing, and collected medicinal roots. Some of the Tubular families had a monopolistic ownership of vast hunting grounds. At the time when the protected area was established, the whole territory east of Chulymshan and the Teletskoye Lake was unpopulated.

The indigenous people developed a code of rules and prohibitions regarding water management. The pollution of running water with waste materials and domestic refuse was forbidden. Medicinal springs – the so called arzhan – were treated

Fig. 1 Location of the Altaiskiy BR (red outline).

with special respect and under protection. Food was prepared in advance before going to an 'arzhan', usually including flatbread, roasted and ground barley, cream and tea. On the way to the springs, no plants were broken or touched, and no hunting or fishing took place. The Altai people respected trees, in particular cedar trees. In the ancient mythology of the Altai people a tree is an elementary object. Just like a mountain peak, a tree represents the centre of the universe and the vertical link between earth and sky. They were aware that natural resources are limited, and only those who use them with care and reason can count on the generosity of nature. The Tubulars and Telenghits, the present inhabitants of the Altaiskiy BR, are still faithful to this belief.

Shift in land-use practices

During the last 300 to 400 years, the Altai Region, formerly populated by nomadic and semi-nomadic tribes with robust family relationships and a pagan religion, found itself at the junction of vigorous penetration by several religions. For centuries, the Altai people were under pressure – from the east and south by the Mongols worshipping Lamaism, and from the west by Kyrghyz tribes who worshipped Islam. From the 17th century onwards, the Russians began to settle in Altai, and as a result, Altai people voluntarily joined the Russian Empire in 1756. This historic event is noteworthy from a geopolitical perspective, as well as from the perspective of the introduction of a different perception of the environment and the development of a new system of land use and resource exploitation. Thus, on the Teletskoye Lake terraces with their favourable climatic conditions, the Russian settlers started planting vegetables and apple orchards that were common in their Slav culture.

In the course of time, the water of the Teletskoye Lake and the rivers and lakes of the Chulyshman Valley were heavily impacted by the shift in land-use. Fishing and hunting were carried out in an unsustainable way. The trophy fauna, such as sable, Siberian weasel, squirrel, Maral Siberian stag, Arkhar mountain sheep and snow leopard, was rigorously exploited and in need of urgent regeneration. Thus, in 1932, by governmental decree, Altai Nature Reserve (zapovednik) was established, with Teletskoye Lake as a strictly protected conservation zone. Boundary check-points were installed to prevent poaching. All types of agricultural land use were prohibited, with the exception of areas along the Chulyshman Valley which were excluded from the reserve in 1936. Nevertheless, in the 20th century, cultivation in the Teletskoye Lake area expanded from small-scale to industrial-scale orchards of over 100 hectares in the economic zone of the nature reserve. In the 1970s, however, it became clear that this gigantic project caused more loss and environmental damage than profit, as there was no market for the apples grown in this area.

The story of Grandfather Alchi

As a result of the difference in land-use traditions, the Tubulars came into conflict with the Russian settlers. The story of Grandfather Alchi tells about a direct confrontation of an indigenous Tubular man, defending his family's cedar against enthusiastic Russian gardeners, ready to clear cedar trees for cultivation:

Grandfather Alchi lived all his life in the Yailu village (translated as 'summer pastures') near the Teletskoye Lake. He was born and raised here. In the middle of the 1960s, the cultivation of apple trees started in Yailu. The land was cleared and trees were felled to give way to apple trees. A cedar was standing amongst others in the foothills of the Torot Mountain. Every year, it produced a rich harvest of nuts which were useful to people as well as animals. When a bulldozer approached the cedar, Grandfather Alchi stepped in raising his hands, ready to protect the tree, and exclaimed 'Come to your senses, people! Look who you destroy! You destroy the feeding hand of all that lives in our taiga woods'. Out of respect for Grandfather Alchi, and for the cedar saved by him, the supervising manager ordered to keep several trees on the meadow as a reminder of the man's deed in defence of the cedar. Ever since, the cedar has been known as 'Grandfather Alchi's sacred cedar'. So, in a way, Grandfather Alchi left a living monument of himself – the sacred cedar. This story from the 1960s reminds us of a Himalayan story of the 1970s, where 27 female peasants in the village of Reni prevented the felling of trees, thus starting the Chipko movement. Chipko supporters sing: *'Maatu hamru, paani hamru, hamra hi chhan yi baun bhi. Pitron na lagai baun, hamunahi ta bachon bhi'*[1]. These words make a robust bridge between the Altai and the Himalayas in the shared desire of indigenous people to protect their environment.

Fig. 2: Listening to the story of Grandfather Alchi's Cedar (in the rear): Graeme Worboys (IUCN), Eugeniy Veselovskiy (Altaiskiy BR), Vasiliy Manishev (Deputy Minister of Natural Resources, Altai Republic) and Yuri Badenkov (Institute of Geography) (© S. Shigreva).

---

[1] The soil is ours, the water is ours, ours are these forests. Our forefathers raised them, it's up to us to protect them (Old Chipko Song, Garhwali language).

New challenges for the Altai region

The 'cedar-apple conflict' of the mid-20th century was finally resolved by reconciling the conflicting systems of using natural resources, practised by indigenous and immigrant people respectively, by adopting a balanced approach based on mutual respect and trust. However, in the 21st century characterised by globalisation and the triumph of the free-market economy, new and acute conflicts on a larger scale have been taking place in the Teletskoye Lake area. These conflicts jeopardise the fragile equilibrium of social and environmental systems. Local and indigenous people wish to raise their standard of living currently based on a rather modest income, by developing environmental tourism and services. Russian and international developers, on the other hand, are willing to invest resources in the development of mass tourism and recreation. Such large-scale investments may undermine local private businesses and could easily destroy the traditions and lifestyle of indigenous and local people. In the late 20th to early 21st century the issue of introducing a development strategy became particularly relevant to the unique Teletskoye Lake area. The situation was aggravated by a decline in the social and economic situation in Russia as a result of the transitional economy, and the Altai Republic was no exception.

Local participation in the Altaiskiy Biosphere Reserve

For quite a significant period in the past, protected areas in Russia were established as 'secluded nature reserves'. As a result, 'zapovedniks' were excluded from regional social and economic development processes, and consequently had no support from the local population. The economic crisis of the 1990s aggravated the conflicts between the jobless population and the strictly protected nature reserves, because the latter denied local and indigenous people access to their traditional resources. Considering all these pending problems, a modern development strategy for the Teletskoye area should allow for a sustainable equilibrium between the three most relevant objectives: the conservation of biodiversity, the support of economic development, and the protection of cultural and historic values. This was the reason why the Altaiskiy BR was established under UNESCO's MAB umbrella in 2009. The zoning regime is meant to guarantee the effective implementation of all three objectives. In this structure, the Altai Nature Zapovednik (the core zone of the territory) adopted the function of leader and coordinator, in line with Soviet/Russian tradition. The objectives of the UNESCO concept are to be achieved by a newly established 'Coordinating Council' cooperating in the management of the Altaiskiy BR. The council involves representatives from various local groups, indigenous Altai communities and the Zapovednik. It is focused on mutually beneficial cooperation between all stakeholders, with the aim to practise sustainable resource management, and to promote a co-ordinated social and economic development policy for the BR's territory.

In 2009, the Altaiskiy BR initiated the establishment of a non-commercial partnership, the 'Teletskoye Lake Board'. Partners in this initiative are: municipal administrations, associations of tour operators, associations of hunters and fishermen, local

Fig. 3: Group of Tubular women at the visitor centre (© S.Shigreva).

businesses etc. The Board plays an important role in the practical implementation of BR principles. This is a new and efficient form of governance in this territory.

The Teletskoye Lake has the additional status of being a UNESCO World Nature Heritage Site. Under these conditions, the Board's objective is to unite conservation efforts and to ensure sustainable development of the adjacent territories. The core principle of the Board's functioning is the involvement of all groups concerned, including the Turochak and Ulagan municipal administrations of the Altai Republic (located within the boundaries of the transition zone), indigenous ethnic Altai communities, tourist businesses and public organisations.

The Teletskoye Lake Board is currently working on a 'Strategy for Sustainable Development and Management of the Teletskoye Area'. At the same time, the Board is in the process of developing plans for the social, environmental, economic, territorial expansion and modernisation of settlements in the Teletskoye area, with the approval of all parties concerned. Future activities include a comprehensive audit of natural resources, of the human potential, the historic and cultural values, as well as the economic, administrative, legislative and all other resources available in the area. Innovative development ideas are to be based on marketing research of current and potential demand and supply in tourism and other services in the Teletskoye area. The development of settlements is to be supported by appropriate spatial planning and architectural design, plans for economic activities, including construction, transport, communications, recycling of waste and agriculture. The Board provides organisational and legal support, internet resources and an information system for the innovative development of the Teletskoye area. The activities listed above may very well reflect the standard routine for many countries and regions. However, this cannot be said of the remote region of the Altaiskiy BR. Recently, local and indigenous population groups have become involved in the management of the resources in their territory. This is a new and challenging situation with regard to the development and protection of the unique social and economic systems of the Altai, and the conservation of its biological and cultural diversity.

Lake Issyk Kul is said to be the second largest high-mountain lake in the world (© Matthias Schmidt).

Central Asia's Blue Pearl:
The Issyk Kul Biosphere Reserve in Kyrgyzstan

by Matthias Schmidt

The post-Soviet states of Central Asia have faced significant changes over the past two decades: The dissolution of the Soviet Union was followed not only by political change but also by the efforts of nation building, economic decline and increasing individual insecurity. To put a region of 43,116 square kilometres inhabited by more than 400,000 people under protection was a courageous idea indeed, considering that the area accounts for more than a fifth of the territory and population of the Kyrgyz Republic which was founded in 1991. The realisation of such a project, however, was probably only possible in view of the fluid times of political and economic transformation, administrative reorganisation and the confrontation with new ideas of nature conservation and strong influences from the West. In 1995 scientists and environmentalists from Germany, supported by the Naturschutzbund Deutschland (NABU) and the Gesellschaft für Technische Zusammenarbeit (GTZ), took the initiative and consulted the Kyrgyz Government to establish a nature reserve zone around Lake Issyk Kul including parts of the Central Tian Shan. In 2001, this region became one of the largest biosphere reserves in the world.

Unique landscape around Issyk Kul
Lake Issyk Kul covers an area of 6,236 square kilometres and is located in the middle of the Asian landmass, in a region dominated by steppes, deserts and mountain ranges. Issyk Kul is said to be the second largest high-mountain lake in the world. Despite its high continental location at an altitude of 1,608 metres above sea level (m.a.s.l.), the (668 metres) deep lake does not freeze up during winter owing to its brackish quality and some hot springs feeding into the lake. One interpretation of the lake's name as 'Hot Lake' relates to this fact. According to another translation, Issyk Kul means 'Holy Lake' which highlights the spiritual and cultural importance of the large lake to the Kyrgyz people. The lake and its surrounding high mountains of the Central Tian Shan, rising up to Pik Pobeda – at 7,439 metres the highest peak of Kyrgyzstan and the second highest of the former Soviet Union – host a unique fauna and flora as well as numerous cultural historical sites (Uhlemann et al. 2003).

Great climatic variations within the region due to its mountainous relief and its immense size are reflected in a variety of ecological zones: deserts, semi-deserts, arid and humid steppes, floodplain areas, coniferous and juniper forests, subalpine and alpine grasslands (Gottschling 2002). Marco Polo sheep (*Ovis ammon polii*), Siberian ibex (*Capra sibirica*) and the endangered snow leopard (*Uncia uncia*) are native. Twelve plant species, eight mammals and fifteen bird species are on the Red List of Kyrgyzstan (Skvortsov 1985). The oligotrophic lake itself hosts only small numbers of fish, a result of fatal experiments in the 1970s when non-native species were introduced that led to the near extinction of local fish species. Apart from its ecological uniqueness and its high number of endemic species, the area around Issyk Kul provides the basis for the livelihood of around 425,000 people (2009). Agriculture in the form of arable farming and animal husbandry has long formed the main economic pillar. Its relevance even increased since the dissolution of the USSR, because many people lost their jobs in the context of economic privatisation processes and now depend more than before on local land and natural resources.

Agrarian utilisation is closely connected to altitude. Settlements, gardens, arable fields and grasslands are located in the vicinity of the lake between 1,600 and 2,200 metres, whereas forests, spring and autumn pastures are found at altitudes between 2,200 and 3,000 metres above see level. The alpine zone ranging from 3,000 to 4,000 metres serves as summer pasture (*jailoo*) for sheep, cattle, horses and yaks (Asykulov & Schmidt 2005). Apart from agriculture, the area around Issyk Kul has valuable mineral resources and is a popular tourist destination. Each summer, up to one million tourists, mainly from Kyrgyzstan itself and from neighbouring Kazakhstan, spend their holidays in one of the resorts on the lake shores.

History of protection
Although most of Kyrgyzstan's population are Muslims, animistic and nature-religious ideas are widespread. Several holy sites (*mazar*) such as springs, hills, mountains, caves or trees can

Fig. 1: Animal husbandry plays a major role in most households in Issyk Kul BR (© Matthias Schmidt).

be found in the area (Dömpke & Musina 2004). They are the destination of many pilgrims who pray at these sites for the realisation of their wishes, such as fertility, rain or healing. As already mentioned Issyk Kul itself is also seen as a holy lake and plays an important role within Kyrgyz tradition and literature, such as in the novels of the world-famous novelist Chinghiz Aitmatov.

Already during the Soviet era, nature conservation zones (*zapovedniki*) were established along the shore line in the western part of the lake and in high mountain areas. These *zapovedniki* followed a strict protection concept according to which people are seen as nature-disturbing factors and should be excluded from the area. Consequently, the *zapovedniki* of the Issyk Kul area covered relatively small, unpopulated and economically unutilised territories. Following the goal of large-scale preservation of the natural and cultural environment around Issyk Kul, new concepts became necessary and were realised in the form of a biosphere reserve (BR). The Issyk Kul BR thus enlarges not only the spatial extent but also the scope of Soviet protection concepts: Humans are explicitly seen as part of the natural and cultural landscape, and its main goals cover not only the conservation of landscapes and cultural sites but also the sustainable economic development of the area. Elements of the development plan comprise the extension of biodiversity, the support of environment-friendly land utilisation practices, including efficient crop rotation, improved irrigation methods and effective usage of fertilisers, as well as ecological education and awareness-raising of the population. Besides, different projects of sustainable agriculture, fruit and wool processing as well as ecotourism were and are carried out (Toktosunov 1998; Hünninghaus 2001).

In line with historically evolved utilisation patterns, the area was divided into four zones: Strict nature conservation obtains in the core zones which are congruent with the already mentioned *zapovedniki* and comprise 3.4 per cent of the territory of the Issyk Kul BR. High mountain steppes and pastures as well as nival regions form the buffer zone that accounts for 81.2 per cent of the whole territory. Arable-field and pasture areas near settlements dominate the transition zones (15.4%), while the reconstruction zones include the main settlements of Karakol, Balykshy and Cholpon Ata as well as the mining areas in which environmental and sustainable reconstruction of old industries and settlements is to be carried out.

Owing to its large size, high number of inhabitants, political instability and economic poverty, the Issyk Kul BR is confronted with tremendous problems. Ecological goals compete with the economic interests of international companies and socio-economic concerns of the local population, while public promotion is weak and the administration is characterised by high fluctuation rates and corruption.

Agriculture and animal husbandry as a livelihood strategy

Kyrgyzstan was one of the least industrialised republics of the former Soviet Union. The agrarian sector always played an important role within the economy of Kyrgyzstan. Almost all households in the rural areas of the country are in some ways involved in agricultural activities. The same holds true for the Issyk Kul region where, according to official statistics, more than 70 per cent of the population sustain their livelihood mainly from agriculture (Statistical Committee of the Kyrgyz Republic 2004). In particular, animal husbandry plays a major role in most households within their livelihood strategies (cf. Fig. 1). Apart from income generation by selling meat, milk products and wool, animals represent an important capital that can be converted to cash when required. After the dissolution of the Soviet Union, the number of livestock and sheep decreased dramatically: In Issyk Kul province, sheep flocks diminished from 1.8 million (1991) to 580,000 (1997), with a slight increase to around 690,000 head in 2009. Additionally, around 175,000 cattle, 75,000 horses and 4,000 yaks are kept nowadays (Statistical Committee of the Kyrgyz Republic 2010). Although the number of animals is much lower than at the end of the 1980s, the productivity of the pastures has declined significantly during the past two decades (Aidarbekova 2007). In particular, easily accessible pastures are overstocked nowadays, because remote pastures are often no longer used. In contrast with the large collective farms (*kolkhozes*) of the Soviet system, today the small agricultural units – mainly households – do not have the personnel and organisational means to send their herds to remote pastures. Generally, using the mountain steppes for pasturing as part of the cultural landscape of the Central Tian Shan, does not conflict with the goals of the BR. However, there might be a problem when the number of grazing animals increases further making their density too high in specific areas, resulting in forest and soil degradation processes. Since other income opportunities are scarce, many people either want to enlarge their private herds (Asykulov 2002) or work as labour migrants in Russia or Kazakhstan (Schmidt & Sagynbekova 2008), while the money they send home is also often invested in livestock (Schoch et al. 2010). Remarkably, the majority of the inhabitants of the Issyk Kul BR do not see any relationship between big herds and negative impacts on nature (Asykulov 2002). However, more than half of the pastures in the Issyk Kul area show clear signs of degradation (Aidarbekova 2007).

Arable farming in the form of rain-fed and irrigated agriculture is practised on seven per cent of the territory of the Issyk Kul BR. Its importance has increased in recent years. Less use of fertilisers and the lack of crop rotation have caused yields to decrease significantly. Furthermore, there is a shortage of agricultural machines and a lack of effective organisation and marketing of agricultural products.

Mineral resources – inheritance burdens and income source

A conflicting discrepancy between economic utilisation and ecological goals is obvious with regard to mining activities in the area. The Kumtor gold mine, run by a Kyrgyz-Canadian syndicate, plays a particularly important role in the national economy of Kyrgyzstan, because it generates more than ten per cent of the national income. The gold mine is located at approx. 4,200 metres above see level in a zone which is extremely fragile in ecological terms (Fig. 2). For the extraction of gold, large amounts of toxic chemicals are necessary, leading to the contamination of glaciers and high mountain steppes in the vicinity of the mine. Tonnes of highly toxic waste are produced day by day and deposited on site. But the mine is located within the catchment area of the Naryn River, the major tributary of the Syr Darya, which is the lifeline for millions of people in Uzbekistan and Kazakhstan. An accident would have dramatic consequences for all Central Asia. Furthermore, the transport of tonnes of fuel and toxic chemicals across the BR represents a constant danger to the environment and a serious threat to the whole ecosystem. Already in 1998, a truck accident led to the spillage of approx. 1.7 tonnes of cyanide into Lake Issyk Kul, resulting in an ecological disaster (Moldogasieva 1998).

Fig. 2: Gold mine in Issyk Kul BR (© Matthias Schmidt).

Fig. 3: The lake Issyk Kul offers many opportunities for leisure activities (© Matthias Schmidt).

Tourism at Issyk Kul

Tourism is another major economic activity in the area. The Issyk Kul region is the setting for various forms of leisure activities, such as water and beach activities, horse riding, cycling and mountaineering, fishing, hunting and wildlife watching (Fig. 3). By far the most important form of tourism is a kind of mass tourism that populates the lake shores during the summer months. Already in the 1970s and 1980s, 27 leisure zones were allocated to the construction of holiday camps and other tourist infrastructure (Lunkin & Lunkina 1987). Thus, most of the tourists numbering almost one million are concentrated in specific local centres on the shores of Issyk Kul, for example in the city of Cholpon Ata. However, the environmental imprint in these locations is significant: The air is polluted by noise and emissions, refuse is produced, pastures and arable land are used for the construction of hotels, holiday houses and recreational camps. Recent developments and plans within the tourism sector do not take the goals of the BR into account. The planned construction of an international airport near the city of Balykshy would improve accessibility and probably increase the influx of tourists thus presumably exacerbating the impact of tourism on the fragile environment. Ecotourism might be an alternative form of ecologically friendly tourism, but so far it has not fulfilled its potential. The tourists' financial means are low, and they are not able or willing to spend extra money on costly environment-friendly services or products. However, it might be a feasible objective, as well as a step forward, to promote eco-friendlier forms of tourist activities, with a clear indication of hiking routes and camping sites combined with well-targeted marketing of local products.

Weak public relations and institutional framework

A major problem of the Issyk Kul BR is the lack of public awareness of its status. The natural beauty and the popularity of the area are obvious. However, the need to behave appropriately in order to preserve this unique environment is not realised by either the tourists, the miners or the local population. Most of the visitors were not aware of the Issyk Kul BR until they entered the territory and paid a small entrance fee. Unfortunately, there are almost no other signs of its unique status within the area. A notable exception is the Ecocentre in Cholpon Ata (Fig. 4). Promotion and marketing of the BR, as well as environment-friendly land utilisation are almost non-existent and there are very few local products available. It must count as a success that nowadays the people of the area are aware at least that they are living within a biosphere reserve (Asykulov 2002), but to most of them the utilisation and protective functions of the various zones are not at all clear. On the one hand, it might be seen as a positive sign that people do not have to change their behaviour much when becoming part of a conservation zone, but on the other hand, carrying on all their economic activities as usual often runs contrary to ecological goals.

The core problem faced by the administration of the Issyk Kul BR is to be found in its complicated political and administrative responsibilities. Land, forests and water are managed by various administrative units of diverse ministries or departments at different regional levels (Aidarbekova 2007). Responsibilities and competencies are not always clear which impedes decision-making processes. In addition, political insecurity and frequent changes in the managerial staff prevent a sustainable policy and development. Nepotism and corruption prevail and lead to a situation in which high positions are held by non-experts.

Conclusion

The establishment of the Issyk Kul BR was a foundation stone and the right measure at the right time on the way to preserving the unique landscape of the Central Tian Shan. The idea of protecting the natural and cultural landscape as well as supporting the development of the local population remains persuasive, but unfortunately, this is far from the present reality in Kyrgyzstan. Political instability, administrative disorder, insufficient financial means, nepotism and corruption as well as severe economic problems stand in the way of compliance with conservation rules and a constructive economic development. The local population needs to use local land and natural resources but is not supported by strong institutions in managing these resources in a sustainable way. In the first instance, their basic needs must be covered and they must have hope for a stable and fair political rule – these aspects go beyond the scope of the Biosphere Reserve.

Fig. 4: Eco-centre in the city of Cholpon Ata (Matthias Schmidt).

References

Aidarbekova, C. (2007). Sustainable management of natural resources in Issyk-Kul Biospheric Territory. Unpublished Report to the European Commission Project 'Development of National Environmental Strategies for Sustainable Development'. Bishkek.

Asykulov, T. (2002). Sozioökonomische und naturräumliche Bedingungen in Ostkyrgyzstan und die Frage über die Entwicklung des Biosphärenreservats 'Ysyk-Köl'. Unpublished PhD thesis, Ernst-Moritz-Arndt-University Greifswald.

Asykulov, T. & M. Schmidt (2005). Naturschutzkonzepte im Transformationsprozess: Das Biosphärenreservat Ysyk-Köl in Kirgistan. Natur und Landschaft 80, 8: 370–377.

Dömpke, S. & D. S. Musina (2004). The call of our ancestors. Natural sacred sites in the Issyk-Köl Biosphere Territory. Bishkek.

Gottschling, H. (2002). Umweltgerechte Landnutzung im Biosphärenreservat Issyk-Kul, Kirgistan. Beiträge aus landschaftökologischer und sozio-ökonomischer Sicht. Heidelberg.

Hünninghaus, A. (2001). Management von Biosphärenreservaten in Transformationsländern. Dargestellt am Beispiel des Biosphärenreservats Issyk-Köl in Kyrgyzstan. Bochum.

Lunkin J. M. & T. V. Lunkina (1987). Touristic zones of Kirgizia. Frunze (in Russian).

Moldogasieva, K. (1998). Ecological catastrophe at Issyk Kul: unexpected scenario and potential consequences. Bishkek (in Russian)

Schmidt, M. & L. Sagynbekova (2008). Past and present migration patterns in Kyrgyzstan. Central Asian Survey 27, 2: 111–127.

Schoch, N., Steimann, B. & S. Thieme (2010). Migration and animal husbandry: competing or complementary livelihood strategies. Evidence from Kyrgyzstan. Natural Resources Forum 34: 211–221.

Skvortsov, V. I. (Ed.) (1985). Red book of the Kyrgyz SSR. Rare and endangered animal and plant species. Frunze. (in Russian)

Statistical Committee of the Kyrgyz Republic (2004). Kyrgyzstan in numbers. Bishkek (in Russian)

Statistical Committee of the Kyrgyz Republic (2010). Census of livestock and poultry in the Kyrgyz Republic in 2009. Bishkek. (in Russian)

Toktosunov, K. (1998). The Biosphere territory and regional development in the Issyk-Kul region of Kyrgyzstan. In: Dömpke, S. & M. Succow (Eds.). Cultural landscapes and nature conservation in Northern Eurasia. Bonn: 200–206.

Uhlemann, K., Vinnik, D. F. & K. I. Ismanova (2003). Biosphärenreservat Issyk-Kul – Inventar der kulturhistorischen Stätten. Eschborn.

Snow clad peaks in the core zone of Nanda Devi Biosphere Reserve (© K.G. Saxena).

Ecotourism in Nanda Devi Biosphere Reserve: A Win-Win Option for Environmental Conservation and Sustainable Livelihoods

By K.G. Saxena, R.K. Maikhuri & K.S. Rao

Environmental conservation in the Himalayas, a mountain system extending across eight Asian countries (Afghanistan, Bangladesh, Bhutan, China, India, Myanmar, Nepal and Pakistan), is crucial with respect to local as well as national and global dimensions of sustainable development. India's recognition as one of the 17 'megadiversity countries' and as one of the ten largest forested areas in the world derives from the Himalayas which, in geographical terms, cover only 18 per cent of the country but account for more than 50 per cent of India's forest cover and for 40 per cent of all species endemic to the Indian subcontinent. India's Protected Areas Network covers around 10 per cent of the Indian Himalaya. It comprises 95 protected areas with a legal status of biosphere reserve (BR), national park or wildlife sanctuary. In the absence of detailed scientific investigation, conservation policies formerly tended to neglect or misunderstand the people-environment relationships which lead to conflicts between people and protected area managers. The current philosophy of the biosphere reserve concept has strengthened the national efforts of avoiding these conflicts and promoting conservation as a means of sustainable rural development. Past experiences and responses in a Protected Area like the Nanda Devi BR have revealed an immense scope for ecotourism. Consequently, sustainable livelihoods of local communities are to be coupled with the conservation and restoration of globally significant biodiversity and environmental services from the Himalayan ecosystems.

The history of conservation and rural development in the Nanda Devi massif

Nanda Devi (7,817 m.a.s.l.) is the second highest mountain in the Indian Himalayas. It forms the summit of a vast glacial basin ringed by high peaks. The interior of this almost inaccessible ring is known as the Nanda Devi Sanctuary. The massif is characterised by an extremely inhospitable climate (temperatures below freezing for five months a year), the highly dissected terrain and poor accessibility. It is fair to say that until 1965, the nearest trafficable road could be reached only by a three-day trek from the most accessible village. This restricted the

human population density in the entire region. Only some indigenous Bhotiya families inhabited the Nanda Devi area. In summer, they grazed their animals, collected medicinal plants and observed religious traditions in the Nanda Devi Sanctuary (nowadays the core conservation zone). The settlements of the Bhotiya have been and still are located outside the sanctuary (which is nowadays the buffer zone of the protected area). Each traditional household moved between two settlements; a large one, where people grew crops in summer, and a smaller one at lower elevations where they lived in winter. Until 1962, the Bhotiyas made a living mainly from bartering wheat, rice and buckwheat with salt and wool from Tibetans in the North; and they bartered wool, high-altitude medicinal plants and crops with the Garhwalis at lower elevations.

Traditional organic food crops – a lesser-known component of biodiversity and tourist attention (© K.G. Saxena).

In the 1970s, mountaineering on the Nanda Devi became popular. The local Bhotiya people were hired as porters and guides for expeditions. By 1977, Nanda Devi had become the second most visited Himalayan peak after Mount Everest. As the environmental impact was growing, in 1982, the Nanda

Building local capacity for promoting ecotourism as a component of reserve management (© K.G. Saxena).

Devi Sanctuary was designated a national park and subsequently closed to all human activities (including mountaineering and grazing). The Bhotiya suffered from this closure because their traditional grazing areas and community forests became off-limits (Bosak 2008).

In 1988, the area was additionally nominated a World Heritage Site and a Biosphere Reserve by UNESCO, with the national park area constituting the core zone and the surrounding area forming the buffer zone. The Nanda Devi BR is the first UNESCO reserve in the Indian Himalaya. After its designation, restrictions on traditional land-use forms were enforced even in the buffer zone. In compensation, the Bhotiya were supplied with solar-powered lamps, improved beehives and spinning as well as weaving devices. A fixed quota of staple food grains and kerosene oil was allocated to each family at subsidised costs. Besides, some income was offered to local people in terms of wages for work in government-funded conservation and rural development projects. In 2000, the neighbouring Valley of Flowers National Park was included in the BR as a secondary core zone. The buffer zone was also expanded so that the entire protected area now covers 5,860 square kilometres. In 2003, the reserve was reopened for tourism which is now regulated by the divisional forest officer of Nanda Devi National Park. Visitors have to pay entrance and overnight fees. The income is shared between the government and the local people. As a rule, only local people can be employed as guides and porters.

Traditional agriculture and indigenous conservation knowledge and practices: a neglected dimension

Agriculture has been a minor land use in terms of spatial extent (1% of the area in the buffer zone) but a highly significant occupation and economic activity. Despite an increase in population, the agricultural land-use area has not expanded over the last 40 years partly because of a legal ban on forest and meadow conversion since 1900 and partly because of indigenous socio-cultural norms discouraging agricultural expansion. For example, practices such as selling any cultivated land to non-indigenous people, or hiring farm labour from outside the village (if a household suffered from labour shortage) have been restrained. Furthermore, according to a religious belief, deforestation for agriculture is certain to invoke catastrophic events. Specific socio-cultural norms allowed for the conservation of the forests while simultaneously maintaining high agricultural productivity. For instance, the privilege of earning income from wild medicinal and aromatic plants, wild food and other non-timber forest products was only given to socio-economically weaker sections of the society (whereas all sections of the society were allowed to utilise natural resources to meet their subsistence requirements). Forest products were only allowed to be used in groups and the exchange of high quality seeds had to take place without any consideration of profit.

Nowadays, people have changed their agricultural practices in response to changing socio-economic and policy factors, such as

- loss of customary cultivation rights in their winter settlements under the formal land tenure policies,
- disruption of trade with the Tibetans after the 1960s due to socio-political factors,
- restrictions on grazing and the utilisation of non-timber forest products following legal protection of the area,
- loss of income following a ban on any expedition and tourism in the core zone enforced in 1980,
- availability of staple food grains at subsidised prices from government ration shops, improvement in accessibility within the lower-elevation region and socio-cultural-policy factors driving the transition from traditional subsistence to a mixed subsistence-market economy.

As a result, local people replaced their traditional staple food crops, such as barley and buckwheat, with cash crops, such as kidney beans, amaranths, potatoes and green peas, demanded at lower elevations and adapted to the ecological conditions in their native high-elevation settlements. However, aspirations for higher income may lead to an irrecoverable loss of less profitable traditional crops even though they might be more resilient to environmental stresses and hence an asset in view of global climate change. Traditional knowledge tends to disappear as people withdraw more and more from livestock husbandry. This is a result of restrictions in terms of access to legally-protected meadows and forests, the improvement in access through trafficable roads, and the emergence of new opportunities of income in government-sponsored projects.

So far, people succeeded in cultivating medicinal species such as *Allium humile, Allium stracheyi, Carum carvi, Pleurospermum angelicoides, Dactylorhiza hatagirea* and *Megacarpea polyandrea*. All agriculture surrounding the reserve is organic and therefore people have a scope of economic benefits from organic food. Many crops and local food dishes still remain less well known to a wider community, especially to the affluent tourist class. Information on such indigenous efforts that are in line with the goals of conservation of biodiversity and natural ecosystems, if disseminated effectively, is apt to attract a new class of tourist (participatory researchers), eco-lovers, enterprising farmers and food technologists.

Building on indigenous concerns about conservation and development

Local people aspire to higher incomes but they have never adopted timber trade as a means of achieving this aspiration, partly because of religious constraints on timber trade and partly because of their perception that forest conversion would lead to a shortage of high-quality fodder and manure and thus unsustainability of the agricultural system, springs drying up, the source of drinking water disappearing and natural disasters resulting from loss of protective cover on steep slopes. A series of policy actions interfering with the traditional natural resource management and land tenure system from 1939 onwards, provoked vociferous public objection only in response to two interventions:

- Local people vehemently opposed any plan of revenue generation from timber trade (not only in the area around the reserve but in all ecologically fragile mountain regions of the country) by adopting 'tree hugging' as an innovative method (you'd have to cut down the people first before you 'd be able to cut the trees), and
- they opposed the ban on mountaineering in Nanda Devi National Park; first they tried to convince government agencies to withdraw the ban; when they found their demand was not properly heard, they defied the ban en-masse and entered the core zone in 1998, with the protection enforcement authorities functioning merely as 'passive onlookers'.

Protected Area planners have rarely recognised the fact that the capital residing in a rich natural resource often derives from the conservation ethos engrained in indigenous knowledge and livelihood systems. This lack of an objective appreciation of indigenous communities consequently led to non-cooperation by or neutral attitudes of local people towards legal protection. People become hostile when they realise that a policy action

- is not backed by systematic scientific studies (e.g. the ban on mountaineering: it is not proven that biodiversity recovered owing to the ban on mountaineering which involved an influx of only 440 tourists and 300 porters over a period of 15 years; the absence of grazing or climate change effects; some people even claim an increase in the frequency of poaching of wild animals and unsustainable harvesting of medicinal plants in the absence of tourists and local people) and
- is aimed at utilising the resources people had conserved for the benefit of 'outsiders' at the expense of threats to their sustainable livelihoods (e.g. revenue generation from timber).

By involving people at all stages – from the determination of a policy intervention to, where necessary, the monitoring of outcomes and adaptations, it should be avoided that policy changes have to be enforced by local people. An example for this is an event that took place in 1974 when locals achieved the adoption of a ban on tree felling; another is the reopening of the core zone to tourists in 2003.

Promoting ecotourism: moving towards a community-centred approach

Since 2003, the core zone of the biosphere reserve has been open to 'regulated tourism', with restrictions on tourist numbers (around 500 tourists per year), entry fees, and employing only local people as porters and guides, avoiding the use of firewood as energy source, disposing of refuse as specified by the forest department and assigning the responsibility for regulation to government agencies. Although local people benefit from mountaineers visiting the core zone, it cannot be ruled out that they will again start raising their voice against regulated tourism in its present form, as there is a

- lack of any systematic participatory scientific study on the carrying capacity in terms of sustainable visitor numbers
- lack of people's participation in deciding on the tourist influx rate and on entry fees and the appropriation of revenue from tourism,
- ambiguity regarding the nature and magnitude of penalties for defying tourism guidelines, and
- limited capacity of the Protected Area management authority for enforcing the adherence to guidelines.

As regulated tourism has been imposed only on the core zone of the Nanda Devi BR, the magnitude of income from tourism and government assistance towards the establishment of new and better tourist facilities may turn into a point of conflict between different local communities around the reserve. It is important to note that traditional agricultural and forest management practices – which tend to strike a balance in the regeneration of resources utilisation and favour equality – could also attract tourists. Unfortunately, such practices have not been incorporated in the tourism development plan. The area can boast great diversity in food crops (approx. 20 species) and there are several local organic food dishes, which, with some modifications, could turn into a popular 'continental menu' and provide people with a higher income at the same time as benefiting the conservation of agro-biodiversity. The goal of sustainable development could be achieved better if protected area management plans were to integrate indigenous and conventional scientific knowledge systems, involve people in the planning as well as in the monitoring of ecotourism, and assign them some of the responsibilities for enforcing regulated ecotourism at the same time as allowing them to reap the resulting economic benefits.

It should be kept in mind that conservation and sustainable utilisation of natural resources are centuries-old inherent dimensions of indigenous culture and livelihood, while policy interventions related to conservation and tourism did not emerge in India until the beginning of the 20th century.

Info-Box on Nanda Devi Biosphere Reserve: Location, people and resources	
Location	• State of Uttrakhand, India; Biogeographic Province Himalayan Highlands • Core zone: 712 km² (comprising Nanda Devi and Valley of Flowers National Park); buffer zone: 5,149 km²
People	• 47 villages, with a total population of 11,000 in the buffer zone • Groups of Bhotiya (an ethnic Tibetan group: Indo-Mongoloid ethnic race) – Jadhs (Buddhism), Tolcha and Marcha (Hinduism), Shaukas (mix of Hinduism and Buddhism) and Garhwalis of Indo-Khasa ethnicity
Ecosystem services	• Huge glaciers feed major river systems, particularly during the dry summer, this is the source of water for millions of people in Asia
Biodiversity	• An extremely wide range of ecosystem diversity: from cold desert ecosystems, temperate, alpine and subalpine forests to alpine meadows and dry scrubs • 400 species of trees, 570 of other plant groups, 534 of birds and mammals, 54 of reptiles and amphibians and 200 of insects, with a large number of rare, threatened and charismatic species (e.g. snow leopard and muskdeer)
Features of tourist attractions	*Physical features:* • A vast glacial basin encircled by 16 peaks including Nanda Devi, the second highest peak (7,817 a.m.s.l.) of the Indian Himalaya • Rishi Ganga river gorge – one of the deepest gorges in the world, with a local relief of 6,000 m • An extremely adventurous trek route in a 'near-pristine' environment • Upper courses of tributaries, with crystal-clear and fast-flowing water, feeding the Ganges river, the lifeline of southern Asia • Glaciers, origin points of rivers and moraines *Biological features:* • A wide range of vegetation types, from mixed temperate forests to alpine meadows, dry scrub vegetation (cold desert) and timberline within the Reserve and across the main trek route • Unique aesthetic value derived from wild flowers across a vast landscape in Valley of Flowers • Rare and charismatic species such as snow leopard, musk deer and many medicinal plant species *Eco-cultural/religious heritage:* • Several globally significant sacred natural objects and shrines (Nanda Devi peak believed to be the living place of the Hindu goddess Nanda Devi and Sikh shrine Hem Kund Sahib and Hindu shrine Badrinath, being the most revered places) and associated cultural functions • The place of origin of the Chipko Movement (tree-hugging) of people that forced the government to ban any green-tree felling in the region in 1974 and subsequently in all fragile hilly regions in the country

References

Bosak, K. (2008). Nature, Conflict and Biodiversity Conservation in the Nanda Devi Biosphere Reserve. Conservation and Society 6(3): 211–224.

Intramontane plain of Arganeraie Biosphere Reserve, Morocco – dotted with argan trees (© M. Boussaid).

Arganeraie Biosphere Reserve, Morocco, and the Role of Women's Cooperatives

by Mohamed Boussaid

Located in the south-west of Morocco, the Arganeraie Biosphere Reserve (BR) covers a huge intramontane plain of more than 2,560,000 hectares, bordered by the High Atlas and Anti-Atlas Mountains and open to the Atlantic in the west. The core protection zone comprises the Souss-Massa National Park. Of main conservation interest is the Argan tree (*Argania spinosa* which is endemic to Morocco. Being a relic of the Tertiary era, this tree species is extremely well adapted to drought and other environmentally challenging conditions (UNESCO 2010). Its deep roots are the most important stabilising factor in the arid ecosystem, providing the final barrier against the encroaching desert. The Argan forests and wooded savannas still cover an area of about 800,000 hectares (Nill & Böhnert 2006). Not only do they act as a buffer against desertification, but they are also a source of livelihood for more than two million people in rural Morocco who depend on the trees for oil, fodder, honey, charcoal, fuel and construction wood. These agroforests suffer from continued degradation induced by intense use such as firewood gathering and grazing in the hilly areas; and, in the plain, from tree removal to introduce irrigated crops (Lacaze 2010).

Threats to the argan ecosystem

The strong traditional links between the communities and the argan tree are the key factor which has guaranteed the stability of this ecosystem over time (Aziki, 2010). In the 20th century, new social and economic changes in the region led to increasing pressure on the argan ecosystem. Firstly, as a result of intensive exploitation of wood to meet the growing needs of the neighbouring cities, secondly by conversion to intensive agricultural production of export crops such as tomatoes. Recently, population growth, resulting urbanisation and the development of infrastructures exacerbated the pressure on the BR area. In addition, the argan forest is exposed to two types of natural hazards: Its geographic location at the front zone of the hottest desert in the globe, and the impacts of climate change on the region (an increase in temperature of 0.6 to 1.1°C, and a precipitation decrease of 10 to 15% by 2020 are expected). Under these new conditions, the supply potential of products and services from

Information on the argan tree

The argan tree belongs to the family of *Sapotacea* (soapwoods). A typical tree has a broad, usually twisted, trunk of up to ten metres in height, with a huge bushy crown, reaching up to 14 metres in diameter. Its deep and wide-reaching root system allow it to make excellent use of the water in the soil. The argan tree can tolerate drought and temperatures in excess of 50 degree Celsius by going dormant. When the first rain falls, it then once again puts out leaves and flowers.

The fruit ripens year-round. In good years, a tree may bear up to three generations of blossoms and fruit at the same time, at completely different stages of maturity. Besides the pulp the fruit comprises an extremely hard nut with two to three kernels, from which a premium oil is pressed. This oil contains over 80 per cent unsaturated fatty acids, Vitamin A, considerable quantities of tocopherol (Vitamin E – antioxidants) and a remarkable quantity of sterols (schottenol and spinasterol). The oil is used as a human food product but since time immemorial it has also been used for skin and hair care, for tending wounds, and against rheumatism and arteriosclerosis (Nill & Böhnert 2006, p.37).

Argan trees in Morocco (© Sylvia Lange).

argan forest is decreasing drastically. The available pastoral resources for grazing and browsing are highly reduced owing to the extended dry season and the increased number of herds. The production of nuts became irregular.

Efforts towards a sustainable use

As part of the 'Programme for Conservation and Development of the Arganeraie' (PCDA), supported by the 'German Technical Cooperation' (GTZ), a framework plan was established between 1997 and 1998, which in December 1998 led to the region being recognised by UNESCO as Arganeraie BR. However, to date there is no formal management team in place for the BR. The Arganeraie Biosphere Reserve Association's Network (RARBA) is playing an important role in conservation (e.g. planting of argan trees), raising awareness and capacity building, including the establishment of cooperatives. The research programme on the argan tree has already been initiated. There are still some unknown requirements in the regeneration of the argan tree (e.g. nursery, transplanting and quantity of watering required). Intensive monitoring and evaluation is necessary in order to assess regeneration success under real conditions. In addition, pilot projects were implemented with regard to natural resources conservation, product enhancement and capacity building. Special support has been given to the development of argan women's cooperatives in the area of the BR.

Currently, there are two techniques in use for extracting the oil: either by hand or mechanically. Hand-pressing is less productive and is practised in the villages. Mechanical pressing achieves a higher level of extraction and greater labour productivity. A prerequisite for this, however, is a regular supply of kernels so that the press can work at full capacity. With this system, processing tends to shift to urban and peri-urban centres, with the rural areas being used purely for the supply of the raw material (Nill & Böhnert 2006, p. 39). The PCDA aimed at supporting the local population, in particular women, in the sustainable use and conservation of argan forests in the BR. One of the direct impacts was the opening up of markets for argan-related products thus increasing the interest of locals and the private sector in the production and trade. The Arganeraie BR provides 20 million days of work for the local population, including some 7.5 million for women's activities, especially oil extraction by traditional techniques (Aziki, 2010. The contribution of argan products to household incomes ranges from 25 to 45 per cent depending on the area concerned. At present, approx. a hundred women's cooperatives – comprising about 3,000 members who represent 500 households – produce about 150,000 litres of argan oil per year. The general oil production by the BR is estimated to about 3,400 tonnes. This business involves more than 30 private companies (Aziki, 2010). In addition, the BR provides about 166 million forage units for grazing animals.

The Tamri women's cooperative

The Tamri forest is considered to be a hotspot of biodiversity in the Arganeraie BR. It harbours the second most important nesting area of the bald ibis, one of the most endangered species

Women of the Tamri cooperative participating in a workshop (© Mohamed Boussaid).

in the area, and hosts other wild fauna and flora. The GTZ, in partnership with the Water and Forest Administration, and the 'projet arganier' (implemented by Social development agency and funded partly by European union) is supporting a women's cooperative in the rural community of Tamri for the sustainable use and conservation of the argan forest. The project aims at involving women in the conservation of natural resources and the value enhancement of argan-related products. The Tamri cooperative includes 84 women and is affiliated to the Union of Argan Women's Cooperatives. Before the creation of the cooperative, the women used laborious traditional techniques to extract the oil from the dry argan fruit. At an individual level, they used to sell their products cheaply in local markets. The generated income did not reward their hard work equitably. The project provided organisational and technical support, such as

- modern machines for oil extraction;
- vocational training for office management and for the members of the cooperative;
- essential fixtures and equipment for a building offered by the rural commune to serve as workplace and sales outlet.

In the course of the project, the cooperative members were pointed to the importance of the argan ecosystem and encouraged to become involved in its sustainable use. In this context, many training sessions were organised covering several issues such as medicinal and aromatic plants and their valorisation using the traditional knowledge which is in general enshrined in women's memories and transmitted from mother to daughter. The cooperative has also been responsible for the reforestation of degraded areas. However, this was a challenging task. In fact, when it comes to the management of collective land, women in Morocco do not have a say. So, the planting of trees in collective areas is a man's decision. Finally, 15 hectares were planted with cactus (*Opuntia ficus indica)* and caroube (*Ceratonia celica)* rather than argan trees in order to avoid conflicts with the agents of the Water and Forest Administration. As the responsibility for land management and use is a sensitive issue, involving property rights, the planting of forest trees remains a controversial issue within the community and requires delicate negotiations. A second component of the project is aimed at supporting the manufacture and marketing of argan products. In addition, measures were taken to ensure product diversification and participation in exhibitions and fairs. In 2006, the argan oil was the first Moroccan product registered as Protected Geographical Indication (PGI) in the EU system in line with the Trademark Protection Law. Proposals for other products (olive oil and goat meat) from Arganeraie BR are currently in preparation.

The arganier project (2003 to 2010) evaluation has shown that in the area of the Arganeraie BR, the number of argan cooperatives has been increasing from 15 in 2003 to 154 in 2010. Even if the women's cooperatives are not directly involved in replanting the argan trees, for reasons outlined above, there was some progress with the reforestation of the area. From 2005 until 2010 approx. 1,300 hectares were replanted with argan trees in partnership with local associations and the High Commissariat of Water and Forest and Combating Desertification.

Conservation of argan forests through commercialisation?

A recent study (Aboudrar et al. 2009) showed that the argan oil boom between 1999 and 2007 had led to increasing pressure on the resources available. Even if people organised in cooperatives tend to protect and conserve their own argan trees, their willingness to conserve the argan forest has not improved. Argan forests are still submitted to over-grazing, wood collection and destructive harvesting methods (i.e. argan almonds are often collected before they are ripe by using sticks). The radical collection of nuts contributes to the problem of a lack of natural regeneration of the forest.

In conclusion it can be said that nature conservation has benefited little from the efforts of establishing women´s cooperatives in the Arganeraie BR. By contrast, the economic situation of the cooperatives has improved, as they now have access to (international) high-value-added markets. Especially in tourist areas, they are able to develop good and promising businesses. Cooperatives with little access to the main tourism market have increasing difficulties in selling their products. As a consequence, these cooperatives are selling their products to other leading cooperatives.

References

Aboudrar, A., Travis, J., Lybbert & Magnan N. (2009). Le marché de l'huile d'argan et son impact sur les ménages et la forêt. Bulletin mensuel d'information et de liaison du PNTTA, ministère de l'agriculture et de la pêche paritime.

Aziki, S. (2010). Etude dans le cadre du projet PRONALCD/GTZ, l'arganeraie du Maroc, un écosystème avec potentialités diverses pour un développement durable.

Benziane M. (1995). Le role socio-économique et environnemental de l'arganier. Acte des journées d'études sur l'arganier, Essaouira, 29–30 septembre 1995.

Délégation de la commission européenne au Maroc (2010): Projet appui a l'amélioration de la situation de l'emploi de la femme rurale et gestion durable de l'arganeraie dans le sud-ouest du Maroc (Projet Arganier). Rapport de mission d'évaluation finale.

GTZ (2001). Synthèse du plan cadre de la réserve de biosphère de l'arganeraie. Projet PCDA.

Lacaze, B. (2010). Monitoring Recent Land-use/Land-cover Changes in Arganeraie Biosphere Reserve, Morocco. Abstract submitted to the '30th EARSeL Symposium: Remote Sensing for Science, Education and Culture', hold in June 2010 at the UNESCO office in Paris. Available at http://www.conferences.earsel.org/abstract/show/1861 (accessed on 21 October 2010).

Nill, D & Böhnert, E. (2006). Value Chains for the Conservation of Biological Diversity for Food and Agriculture. Potatoes in the Andes, Ethiopian Coffee, Argan Oil from Morocco and Grasscutters in West Africa. Published by GTZ and GFU.

UNESCO (2010). MAB Biosphere Reserves Directory, Internet portal at http://www.unesco.org/mabdb/br/brdir/directory/biores.asp?mode=all&code=MOR+01 (accessed on 16 October 2010).

Traditional coffee ceremony in Kafa Biosphere Reserve, Ethiopia (© Svane Bender-Kaphengst).

Conserving Biodiversity for the Sake of Local People: Why Biosphere Reserves are Ideal Development Instruments for Ethiopia

by Sisay Nune

Overview of the problem

Various studies agreed that the natural forest cover in Ethiopia used to amount to some 40 per cent. Owing to a variety of causes, this cover dwindled to less than 4 per cent (WBISPP, 2004). The associated ecosystem services declined more and more. While the forest cover and wildlife areas of the country were lush, people in rural communities had a variety of options to gain their daily livelihood. To mention just a few: harvesting wild fruits and hunting. And crop yields were greater because of conducive microclimates. Rivers ran dry before people noticed and fish disappeared. The scarcity of wood fuel forced rural communities to burn cattle dung and crop residue as substitutes. As a result, soil fertility went into a decline. The search for more land continued: A few steep slopes and riversides came under cultivation. Later the encroachment on this land also intensified. Moreover, the population of Ethiopia has increased from time to time, and a recent official census has revealed that the population currently numbers more than 79 million people (CSA, 2010). Poor environmental policy and management can lead to serious environmental degradation, even in the absence of population growth. Policies that focus on the large-scale development of crops such as palm oil, sugar cane or rice are known to be drivers of deforestation and displacement of local communities. Lack of knowledge promotes the creation of weak policies.

Biodiversity guarantees the effective functioning of ecosystems and contributes to reducing poverty, ensuring food security and human health (Bird 2008). For developing countries such as Ethiopia, biodiversity makes a major contribution to the welfare of rural communities; more than 80 per cent of the population relies on traditional medicines to cure a variety of diseases that affect both humans and animals. Natural honey production would be impossible without availability of the pollen source. Rural communities produce honey in great quantities. The pollen source is mostly to be found in the natural vegetation. Studies have argued that even agricultural crops need cross-pollination with natural vegetation to produce greater yields and higher-quality crops.

Areas with less vegetation tend to dry out more easily than areas with better vegetation cover. In this case, productivity per unit area will be constrained by the availability of water while other crop-production factors remain the same. Areas with better vegetation cover meet the local communities' demand for products and services. Spices, condiments, mushrooms, climbers for house construction (in lieu of nails), firewood (the main and only source of energy), are among many natural products harvested by local communities.

On the other hand, the expansion of urban areas coupled with population growth changed both the attitude towards and the amount of food consumption. As a result of these phenomena, production and exploitation intensified. In recent decades, approximately 20 per cent of the world's freshwater fish have become extinct, threatened or endangered, while roughly 75 per cent of the major marine fish stocks are either depleted, overexploited or being fished to capacity (Bird 2008). The pressure to increase food production means, for example, that more and more virgin land and forests are transformed for agricultural use, increasing environmental problems such as erosion and soil depletion.

The Earth Summit in Rio 1992 recognised that 'the major cause of the continued deterioration of the global environment is the unsustainable pattern of production and consumption, particularly in industrial countries, aggravating poverty and imbalance' (UNCED Agenda 21). Nevertheless, the situation is aggravated to this day by various mechanisms such as Foreign Direct Investment (FDI) in agriculture. FDI is damaging the environment and creating social coercion. Evidence has been presented in various media reports and studies showing how FDI affects the stability of a country. In its policy brief, IFPRI indicated (April 2009) that in Madagascar, the Daewoo Logistic Cooperation leasing 1.3 million ha for maize and palm oil production reportedly played a role in the political conflicts that led to the overthrow of the government in 2009. Globally, 20 per cent of the world's population in the highest-income countries account for 86 per cent of total private consumption. The impact of consumption on the environment therefore cannot be seen in isolation from the problems of poverty, health and quality of life affecting many of the world's nations. The fact is that current consumption is affecting the biodiversity of the world, including the African continent and Ethiopia. Linked to the problem of over-consumption are the issues of inequality and poverty. Almost 1.3 billion people worldwide live on less than a dollar a day.

The biosphere reserve context – how do we understand it?
After various tools had been tested in the field to 'save our planet' UNSECO developed an innovative tool which can solve the problems of conservation and development. The tool emphasises that conservation and human development should go together. Development should not come at the expense of the degradation of natural resources. Otherwise the concept of 'sustainable development' simply does not apply and the fate of the human race on this planet may be its extinction. Initially, however, the depletion of natural resources affects the poor and the immediate user groups of these resources. The UNESCO concept of biosphere reserves requires the full participation and resolve of local communities, local governments and all stakeholders at various stages. This enables a local community to have ownership rights over their resources and to share the responsibility of developing, utilising and conserving its resources. The ownership rights increase their confidence that any outside development will not evict them from their area. This is in line with Ethiopia's constitution. A biosphere reserve naturally requires well-functioning and representative eco-regions where various species of plants and animals occur. A biosphere reserve contributes towards protecting a number of threatened types of vegetation or 'hotspots'. Those hotspots can be havens for avifaunal biodiversity, or they can provide habitats for a great number of birds. Wetlands that serve as a headwater region for most of the rivers not only sustain millions of people downstream but also recharges the hydrological cycle. Such headwaters also sustain forest habitats without which a number of bird species, for example, cannot exist. And they play an important role in traditional forms of agriculture or land use, where either subsistence farming or commercial agriculture are developed by a combination of small holders and medium to large-scale farmers. The tool would provide protection for a number of amphibians (which occur in marsh areas), invertebrates and reptiles. It would also contribute to the preservation of an ever-decreasing habitat for endemic and globally threatened wildlife, and it would contribute to the rehabilitation and appropriate management of river systems and streams which – despite having been modified – are extremely worthy of conservation. Actually, the mosaics of ecosystems mentioned above, combined with various degrees of beneficial human intervention offer the only chance of combating climate change – the most serious threat to humankind in this century.

A biosphere reserve would contribute to the protection of the unique cultural, historical, geological and aesthetic qualities of the country. In this context it must be remembered that the cultural landscape is characterised by a fusion of different cultures and their historic development. In Ethiopia there are unique ceremonies among which the wedding ceremony and the funeral ceremony are worth mentioning, as in fact are other cultural events when local communities come together to worship nature in a variety of languages and a panoply of rituals (dance, gestures etc).

Zooming in on the physiographic nature of Ethiopia
Elevations range from 170 meters below sea level in the Danakil Depression to 4,542 metres above sea level at Ras Dashen, Ethiopia's highest mountain (IFPRI 2006). The Central Statistics Agency has listed 58 mountains in the country which have an elevation ranging from 2,989 to 4,542 metres (CSA, 2010). Almost 44 per cent of the land mass is in the highland area at altitudes above 1,500 metres above see level, which carries more than 90 per cent of the overall population (EFAP, 1994). The FAO (Food and Agriculture Organisation) estimates that approx. 50 per cent of African mountains, i.e. a land surface of approx. 371,432 square kilometres above an altitude of 2,000 metres, are located within Ethiopia (FAO 1984). The lowland areas also exhibit rugged terrains although there are plains in some areas.

Erosion poses a major threat to agricultural land use in Ethiopia (© Sisay Nune).

Because of the topography coupled with long years of poor cultivation methods as well as population pressure, the highland ecosystems are almost exhausted. In 1994, the Ethiopian Forestry Action Programme (EFAP) named three main causes of land degradation which left most of the highlands bare of vegetation: a) population growth, b) low agricultural productivity, and c) high dependency on wood fuel as a source of household energy. Consequently,

- More than 14 million hectares in the highlands were seriously eroded, 13 million hectares were moderately eroded and of the remaining 28 million hectares, 15 million hectares were susceptible to erosion (FAO, 1985);
- Approx. 1,900 million tonnes of soil were eroded annually from the highlands (1985).

The government of Ethiopia and Donors have been spending billions of dollars on food security. Since 1974 the government has been receiving grain from developed countries. Every year there is an increase in the amount of aid and the number of aid-assisted people. In the recent past the government and donor agencies agreed to invest into the creation of assets through various interventions among which soil and water conservation activities, building schools, access roads, are worth mentioning. This effort was and still is costly. The efforts mainly target soil and water conservation while saving millions of lives on a temporary basis. However, these efforts are reactive rather than proactive. On top of that, the lowlands have numerous problems associated with desertification and water stress.

Why biosphere reserve is considered as an option now

One of the main reasons for the degradation of land in Ethiopia can be attributed to lack of proper planning and coordination among various offices, in addition to the other factors mentioned above. The Ministry of Agriculture's main aim is to increase production, no matter where it takes place. The Ministry was under-resourced in terms of the finance and skills required to manage natural resources adequately. The agricultural sector is in constant competition with the natural-resources sector. Exploitation therefore exceeds the regenerative capacity of natural ecosystems.

A biosphere reserve calls for collaboration and cooperation in the interest of sustainable development. It requires coordination among all parties concerned. The Madrid Action Plan calls for increased cooperation and coordination of biosphere reserves with existing international, national and state programmes and initiatives in order to work closely with the authorities responsible for the implementation of relevant biodiversity and environmental multilateral agreements to ensure coordination with other biodiversity. Land use conflict is one of the problems for development in general. Large areas of biodiversity hotspots are given over to agriculture and settlement without due consideration of the potential consequences. Land-use conflicts can be over-come by working in partnership on joint work programmes which match both the needs of the local people and the site itself, with various complementary activities undertaken. Stake-

holders and collaborators get together to discuss and make joint decisions regarding human development bearing in mind the conservation issues of the area in question. The allocation of land will be more productive if it takes place with the benefit of spatial planning in order to achieve environmental, social and economic sustainability. An appreciation and acknowledgement of the indigenous knowledge of natural resources as well as the social capital can support efforts made within national and international agendas, such as the Millennium Development Goals, with the aim to achieve good integration of people and wildlife. The most important thing probably is to draw lessons from the important social capital which has contributed to the conservation of the existing biodiversity and to apply those insights to other areas of the country and to the continent as a whole.

Biosphere reserves provide equal opportunities for conservation and development on the strength of their zonation concept of Core, Buffer and Transition Zones. Zonation clearly strikes a balance between the promotion of human development and the conservation of biological diversity (Ethiopia has experienced severe losses in biological diversity. At the same time, countless human and animal lives have been lost or at least affected). Thanks to its zonation concept the BR tool contributes to the conservation of landscape, ecosystems, species and genes. For example, apart from its value to the dynamic health of ecosystems, the conservation of *Coffea arabica* is of great economic value. Many industries in the country as well as internationally depend on the continued supply of coffee from rural parts of coffee growing regions. Conservation of the species guarantees the smooth and sustainable supply of coffee products. Culturally and ecologically sustainable development is the second most important outcome that is expected from biosphere reserves.

Harvesting wild coffee is a major source of income which does not harm the natural forests (© Svane Bender-Kaphengst).

The other important function is that research, monitoring and education take place. These three functions are linked spatially through the concept of zonation. Most of the land in Ethiopia, particularly in the highlands, is mountainous and of great scenic value if well preserved. But cultivation and settlement have affected most of these beautiful mountains. Education programmes within a BR context can help to change the attitudes of people in local communities.

Through careful planning, biosphere reserves can provide alternative means of income generation such as eco-jobs, eco-tourism and marketing of various non-timber forest products. A biosphere reserve brand is of great economic value. It can help to decrease a local community's dependence on agriculture. Formal education provides very limited courses on environmental subjects and natural resources. In most cases the local population is unable to read or write. The logistic support of a biosphere reserve will enable the local community to visualise the future both with and without their natural environment and will enable them to decide whether they want to continue living in harmony with nature or not. Educating the local people on various aspects of human development and conservation at the same time as developing strategies for long-term cooperation and partnership can strengthen their capacity to manage their natural environment. For a country like Ethiopia, one of the constraints to natural-resources management is sustainable finance. If it were possible to achieve long-term cooperation and collaboration this would help the government not to spend limited financial resources on developing and implementing legal regulations which could instead be invested in activities that are planned better and more wisely.

Ethiopia is signatory to many of the international conventions. The country has a limited number of endemic species of wildlife including birds. There are also indigenous and threatened plant species in the country. The conservation of the diversity of flora and fauna will allow Ethiopia to meet its international commitments. BRs therefore play a vital role in the conservation of such valuable biodiversity at local, regional and international level. By doing so Ethiopia is true to one of the internationally agreed mottos, *'Think globally and act locally'*. Actually, when a country dedicates an area to a BR, this benefits society as a whole as well as complying with Agenda 21 of the 1992 UNCED.

A biosphere reserve is based on the premise that development can serve as a primary economic driver which unlocks funds to support, in a meaningful and sustainable manner, economic growth, social development and environmental rehabilitation. Key requirements are that the economy prospers and that the efficiency of state and NGO spending be increased. Such efficiency can be considerably enhanced through focused and much needed public-private-community partnerships. Private-sector involvement complements conservation and development through making rural-urban links. Without private-sector involvement, it is not easy to market local products. Selling various products from a biosphere reserve can increase

a household's income. Infrastructure development such as eco-friendly lodges, feeder roads and hotels is more easily achieved, if private-sector involvement is encouraged.

People are central to a biosphere reserve. Local people have rights and responsibilities regarding the resources in their area. Power to determine the use of their resources indirectly increases their overall confidence that their environment is well protected for their own sake. In a way, environmental governance is improved when it becomes impossible to evict members of the community from their holdings as a result of allocating large agricultural areas to commercial farming. Local people will not be displaced, natural resources will not be depleted, and no chemicals hazardous to the environment will be used. An environment with good conservation status provides various options that can alleviate poverty. In the face of climate change, water scarcity is one big problem for most African countries. Despite reports indicating the availability of annual runoff volumes of 122 billion m^3 of water and an estimated 2.6–6.5 billion m^3 of groundwater potential, Ethiopia suffers every year from lack of sufficient and timely rains. Only a fraction of the total amount of runoff indicated above remains in the country. At the same time, high surface runoff does a great deal of damage to the infrastructure and sometimes leads to inundation of residential areas from which most of the inhabitants are displaced. If a watershed is located in a biosphere reserve, spatial planning regulations will ensure close monitoring of steep slopes and de-vegetated areas and soil and water conservation measures are implemented; for example, biological and physical conservation methods are applied. In this way, water conservation provides sustainable water flow throughout the year, so that local communities and neighbouring countries have access to water while, at same time, rivers provide fish for local consumption. It is possible to employ small-scale irrigation in the transitional zone of a biosphere reserve, which can give the local population opportunities to grow various agricultural products thus increasing their annual income. At the same time, controlling environmental damage means decreasing the state's investment on disaster prevention and rehabilitation.

Currently climate change is high on the political agenda. The Prime Minister of Ethiopia is very keen to combat climate change. Various ministries are also dealing with climate change. Biosphere reserves may be the only tool that can fully address adaptation to climate change and mitigation issues with regard to the current rural development strategies of the country. Ethiopia started by dedicating two important areas to the UNESCO MAB Programme. There is also a strong tendency to dedicate other areas of the country to the same purpose. But the effectiveness of (the existing) BRs is going to be monitored closely by all stakeholders including the international community. The better a biosphere reserve performs, the more likely the designation of other biosphere reserves.

The Information Centre is where knowledge is managed in a biosphere reserve. Information from various 'Man and the

Biosphere' networks are exchanged and documented. And best practices are disseminated by the Information Centre. The information can guide decision-makers in designing better informed policies.

Among other things, sustainable development safeguards peace and stability. There is a strong negative correlation between conflict and human development: According to Principle 25 of the Rio Declaration on Environment and Development (UN 1992): *'Peace, development and environmental protection are interdependent and indivisible.'* Various media reports indicate that tribal conflict over resources such as water and grazing areas is quite common in Africa. African leaders are well aware of such problems and a lot of resources are spent every year in the name of conflict resolution. The ultimate goal of a biosphere reserve is to bring sustainable development to the region.

As the resources available to people are diminishing – for example, through the loss of access to land and other natural resources on which livelihoods depend, and the loss of access to education and health care – and so is their freedom of choice. At such times men and women, children and youth migrate to major cities or to nearby towns, thus creating social pressure on various resources such as water, food and shelter. Crime and prostitution also increase. Biosphere reserves can be one of the best places where equitable sharing of resources takes place and access to the available resource is granted with equal opportunity for all. In this way, resources are not diminished at all. In fact, optimised use of resources enables sustainable productivity. Everybody who lives within the biosphere reserve has equal access to all the available resources and benefits.

Conclusion

A biosphere reserve provides the opportunity to work together locally, nationally and internationally. It is a mechanism for implementing the United Nations Convention on Biological Diversity in order to achieve its objectives: conservation of biological diversity; sustainable use of its components; and fair and equitable sharing of benefits arising from the utilisation of genetic resources. Ethiopia is a mountainous country. The country has for some time been affected by climate-change related phenomena. Currently, investment is made in safety-net programmes. The Sustainable Land Management programme (mainly soil and water conservation), the World Food Programme, conflict resolution mechanisms and capital-intensive land management practices do not have long-term sustainability as they all lack coordination, collaboration and empowerment of local people. They therefore lack the indigenous knowledge to manage natural resources, as they are characterised by donor-driven technologies. A biosphere reserve is a laboratory and, at the same time, a model for sustainable development. The available natural and environmental capital within a biosphere reserve will sustainably enrich and develop the human capital. In order to achieve this, it is an essential prerequisite to create or enhance the social capital. Building trust is a key requirement for sustainable development. A biosphere reserve empowers the local people who care for and look after their social capital. Ethiopia's land management will be much improved, if more biosphere reserves can be established in this country.

References

Bird, E., Lutz, R., and Warwick, C. (2008). Media as Partners in Education for Sustainable Development: A training and Resource Kit. United Nations Educational, Scientific and Cultural Organisation, Paris, France.

v. Braun, J. & Ruth Meinzen-Dick, R. (2009). 'Land Grabbing' by Foreign Investors in Developing Countries: Risks and Opportunities. IFPRI Policy Brief 13, April 2009. (Available at: www.ifpri.org/sites/default/files/publications/bp013all.pdf)

CSA (2010). Internet portal of the Central Statistical Agency of Ethiopia at http://www.csa.gov.et/

Ethiopian Forestry Action Program (EFAP) (1994). Ministry of Natural resources Development and Environmental Protection. Addis Ababa.

FAO (1985). Ethiopian Highland Reclamation Study. Rome.

IFPRI (2006). Atlas of the Ethiopian Rural Economy. Addis Ababa, Ethiopia.

WBISPP (2004). A National Strategic Plan for the Biomass Energy Sector. Addis Ababa.

The cloud forests in Kafa Biosphere Reserve contain wild Arabica coffee plants (© Bruno D`Amicis).

Saving the Wild Coffee Forests
Joint Forces for Kafa Biosphere Reserve in Ethiopia

by Svane Bender-Kaphengst

Summary

Ethiopia is one of the most fascinating countries in the world, but also one of the poorest. It offers impressive landscapes and unique biological diversity. However, it is facing an enormous growth in population, which is leading to a consumption of natural resources that is no longer sustainable. The highland rainforests of the south-western plateau of Ethiopia are considered to be the origin of Arabica coffee and still harbour many wild coffee varieties – an invaluable genetic resource. But, due to deforestation, the diversity of what is estimated to be approx. 5,000 varieties is in danger of being irretrievably lost. The establishment of a UNESCO Biosphere Reserve has given us the opportunity to merge both the preservation of the remaining unique coffee forests and the sustainable development of the region. NABU, the German Nature and Biodiversity Conservation Union, has supported and backed the Ethiopian government over a number of years to realise the idea. Only two and a half years later, in June 2010, the area was officially designated by UNESCO as Kafa Biosphere Reserve (BR) – an inspiring success. The Kafa BR is one of Ethiopia's first two biosphere reserves and the first coffee biosphere reserve in the world – an attraction for coffee lovers worldwide.

Kafa – a place of outstanding value

The mysterious wild coffee forests are situated in the south-west of Ethiopia, in the Kafa Zone. This zone is located in the Southern Nations, Nationalities and People's Regional State (SNNPRS), the most ethnically and linguistically diverse part of Ethiopia. The predominantly highland region is covered with evergreen montane forest and is part of the Eastern Afro-montane Biodiversity Hotspot. The plateau, which was formerly densely forested, still has primeval forests, bamboo thickets and wetlands and is home to the wild-growing *Coffea arabica*. In the forest, giant trees, lianas, epiphytes and ferns form dense, green vegetation that harbours an abundance of plants and animal species, including the striking black and white colobus monkey. Lions, leopards, wild cats, De Braza's monkey, bush pigs and antelopes such as the red forest duiker and the hartebeest roam the forests. According to several ornithological studies, approx. 260 bird species have been listed in the area, qualifying it to be registered as an Important Bird Area (IBA). The numerous wetlands and the three major rivers, the Gojeb, Dinchia and Woshi, make the forests an important fresh-water reservoir.

Unique wild coffee forests

Around 90 per cent of the coffee drunk worldwide is Arabica coffee. The cloud forests in the Kafa region form the habitat of the last remaining populations of wild-growing *Coffea arabica*, and are considered to be the original source of this species. Scientists estimate that in Kafa centuries of mostly undisturbed evolution have produced around 5,000 varieties of coffee. Coffee plants are a part of the delicately balanced forest eco-system in Kafa and are used by the local inhabitants. The coffee is picked both for personal use and for sale at local markets. A typical farmer still lives on what is grown in his fields and harvests the wild-growing coffee fruit and a variety of commercially-valuable spices and honey from wild bees for his own use and sale at local markets. Nowadays, over 6,500 farmers have formed cooperatives through which they can supply more coffee at a consistently high quality than they could as individual farmers. Now the coffee from the cooperatives is even exported internationally.

It's full speed ahead for Kafa Biosphere Reserve

NABU was asked in 2006 to join a German Public-Private Partnership (PPP) project with a number of private companies, NGOs and the German Technical Cooperation, each offering a different, yet complementary, range of skills and expertise. Amongst the partners for the model project for sustainable development and forest conservation were Geo Rainforest Conservation (GEO), German Foundation for World Population (DSW), German Technical Cooperation (GTZ), Kraft Foods and Original Food. Thanks to its expertise in the management of protected areas on an international level, NABU encouraged and supervised the development of the UNESCO Biosphere Reserve in Kafa. The concept opened up new opportunities for the region and for the country as a whole: Untouched core zones of nature, surrounding buffer zones and a large transition zone, would offer room for conservation, research and development. A large-scale biosphere reserve would be able to increase the population's income through the export of wild coffee to Europe and provide additional marketing opportunities both for local products and for tourism to the 'birthplace' of coffee. The existing participatory forest management scheme (PFM) and family planning programme could be extended and improved. When the concept was presented in Ethiopia, governments at local, regional and national levels welcomed the approach and offered generous support. After official consultation at regional and community level, planning workshops were held and government staff were trained. Subsequently, 'demarcation committees' were nominated and time-consuming resource mapping with the local communities involved was conducted. Once all stakeholders had agreed upon a zoning scheme, it was possible to start the actual demarcation work. Incredibly, more than 500 representatives of the region took part in the process of zoning the biosphere reserve area with the aim of establishing an appropriate management scheme and ensuring the protection of the forests. The initial idea for the establishment of the Kafa BR was raised by NABU in the framework of the German Public-Private Partnership Project. Before long, the Zone's government and the majority of the local population warmed to the idea of establishing a protected area. Both politicians and locals were optimistic about the opportunity for the region to raise its profile by means of the UNESCO's emblem and to foster regional development by selling the wild coffee under the Kafa BR's brand. Another motivating factor was the official training of government staff and community representatives for the zoning process. In the communities, people were continuously kept informed of developments which gave them an understanding of the concept and raised their confidence. The increased confidence created numerous committed multipliers and induced people to become community representatives for further activities. However, this motivating success could never have been realised without support from the excellent Ethiopian staff.

Nevertheless, some of the traditional forest users suspected changes to their customary usufruct arrangements, and the Ethiopian investment agencies as well as private investors in coffee plantations raised their voice against the plan. In addition, there were challenges during the work on the ground: rainy and cloudy weather hindered GPS work on demarcation. There was a lack of expertise in digital geographic data processing, and local government representatives did not always succeed in obtaining correct relevant data. The process was slowed down by general shortage of time, in particular of administrative staff; not to mention newly emerging arguments with adjacent farmland owners in the course of demarcating the boundaries.

Over a number of years, Ethiopian and international experts have collected geographical and scientific data and implemented research projects on flora and fauna, land use, management of

More than 500 people supported the zoning of the Kafa Biosphere Reserve (© Svane Bender-Kaphengst).

natural resources and tourism. A comprehensive management plan for the Kafa BR was established and will be implemented step by step. The official management body will be affiliated to the Kafa Zone's Department of Agriculture & Rural Development in Bonga town and its related administrative offices in the countryside. In September 2009, less than three years after NABU's first attempts to present UNESCO's concept to the Ethiopian government, the application was submitted to UNESCO in Paris by the Ethiopian Ministry of Science & Technology.

Natural and cultural assets of the Kafa Biosphere Reserve

The Kafa BR is characterised by impressive natural scenery which extends over more than 760,000 hectares. Lush ancient forests, thriving wetlands, steep valleys, towering mountains, and gently rolling plains invite the visitor to venture further. The range of altitudes creates a transition of flora: At the highest altitudes, a complex vegetation structure of evergreen mountain forests and grasslands is dominant, while further down the mountain slopes, the Afromontane moist evergreen broadleaf forest or cloud forest is home to the wild *Coffea arabica*. When reaching the lowlands, the visitor encounters woodlands and gently rolling hills. The reserve's capital is the town of Bonga surrounded by Kafa's forests, hot springs, caves and waterfalls. Like other parts of Ethiopia, the Kafa BR is seismically active and contains awe-inspiring hot springs. These waters are recognised for their spirituality and curative value. Hiking to these spiritual places where also local people come to bathe at weekends is an attractive activity for tourists, especially as exciting plants and wildlife can be seen on the path winding through the cloud forests. Numerous natural cave formations are scattered throughout the Kafa BR, formed by underwater streams carving the soft limestone into impressive caverns. Most of these caves are found in the cliffs near springs or rivers and in the thick forests and are shelter for wildlife such as bats and a number of mammals. Visitors can also visit a famous cave from which numerous hyenas emerge at dawn.

As well as nature, a fusion of past and present spirituality, as well as the remoteness and local traditions will fascinate the visitor. The last remnants of the great Kafa Kingdom, like the ingenious defence trenches and watch towers that extend along most of the borders of the former kingdom, can be visited in the Kafa BR. The palaces of Andracha and Sherada of the Kafa Kings were unfortunately demolished by the troops of Emperor Menelik II, but will be reconstructed and brought back to life in an open-air museum. Nevertheless, the main attraction of the Kafa BR is the coffee culture. Ethiopia is the only coffee-producing country in Africa with a traditional coffee-drinking culture. At least three times a day, women perform a ritual, the daily 'coffee ceremony', when green coffee beans are freshly roasted, crushed and brewed to be served to family and guests sitting together to discuss events and share stories. After decades of research, Kafa's profile as the 'birthplace' of Arabica coffee was raised recently when the Ethiopian government decided to establish the National Coffee Museum in Bonga, the Zone's capital.

Traditional land use shapes the biosphere reserve

Apart from collecting wild coffee in the forest, the local population grow 'garden' coffee and farm crops such as barley, teff, sorghum and maize as well as vegetables and fruit. Like everywhere in Ethiopia, livestock plays an important role, and, according to their wealth, families own a smaller or larger herd of cattle. Most of the people living within the BR rely on the environment and its resources for their subsistence. The major source of livelihood is traditional agriculture, along-side livestock rearing and the collection of Non-Timber Forest Products (NTFP). The forests in the Kafa BR are an important source of coffee, useful fruit, medicine, spices, honey, beeswax and timber products such as firewood, charcoal, bamboo, lianas and other building materials. Some 657,780 people currently live and work in the BR (see Tab. 1). The overall majority of the people

The story behind the Kafa Forest Kingdom

In 1879, the last protector of the Kafa Kingdom, a priest chosen by the divine Kafa king, fought his way through the dense forests, desperately looking for a safe haven for the king's possessions. As long as the king's regalia – the crown, bangle and other symbols of divine power – stayed in Kafa, the kingdom in the wild coffee forests could survive. After months of being chased relentlessly, he finally succumbed to his pursuers and failed in his mission. Only 130 years ago, the Kafa Kingdom was over-thrown by the Abyssinian Emperor Menelik II which resulted in significant loss of life, destruction of buildings, displacement of the population and enslavement. Since the mid-14th century, the area in south-western Ethiopia has been described as having mostly dense cloud forests which contain wild *Coffea arabica*, enclosed by earth walls and thus strictly protected from any intrusion The Kafa Kingdom accumulated its wealth through trade in gold, ivory and coffee with northern Ethiopian empires and by means of tribute paid from smaller kingdoms in the surrounding area.

(91.68%) live in rural areas, whilst only about 8.42 per cent live in urban areas, with Bonga, Wacha and Shishinda being the largest settlements. In terms of the average age of the population, it is best described as relatively youthful, with approximately 44 per cent of the population aged 14 years and younger. People live in traditional round clay huts called 'tukul' in villages and hamlets scattered across the countryside.

Steps towards sustainable development of the Kafa region

In 2009, NABU initiated a four-year project on 'Climate Protection and Preservation of Primary Forests' in the Kafa BR. Funded by the German Federal Ministry for the

Location	Kafa Zone in Southern Nations, Nationalities, and Peoples Regional State (SNNPRS), Ethiopia
Total extension	760,144 ha
Core zone	41,391 ha (including 11 National Protected Forest Areas)
Candidate core zone	219,441 ha (no statutory conservation status)
Buffer zone	161,427 ha
Transition zone	337,885 ha
Area covered in forest	422,260 ha (=55.55% of the total surface)
Major ecosystems and habitats	Sub-Afroalpine habitats with moist evergreen montane cloud forest containing wild *Coffea arabica*, bamboo thickets and grasslands Combretum-Terminalia bushlands, aquatic habitats (river systems, wetlands)
Population	657,780 (44% aged 14 and younger)
Ethnic composition	Kafecho (81.4%), Amhara (5.5%), Oromo (2.35%), others and indigenous groups such as Manja (5.38%)
Persons per household	4.4
Population growth rate	2.9%

Table 1: Statistics for the Kafa Biosphere Reserve in Ethiopia.

Environment, Nature Conservation and Nuclear Safety (BMU) within the framework of the International Climate Initiative, the project supports the reforestation of 700 hectares of natural forest with native tree species and the planting of 1,500 hectares of fast-growing trees in community forests next to villages to ensure the population's wood supply. Furthermore, 10,000 wood-saving stoves are introduced in selected communities to reduce the communities' reliance on forest resources. About 10,000 hectares of natural forest will be jointly identified by the Ethiopian Government of the Kafa Zone and the Kafa BR Management following the principles of sustainable Participatory Forest Management (PFM). Tourist infrastructure such as hiking trails, wildlife, and bird-watching hides and a historical outdoor museum are to be developed, and locals will be trained as guides. The proposed scheme will make a significant contribution to the preservation of biological diversity. The development of tourism, creation of jobs and a microcredit scheme within the boundaries of Kafa BR will significantly improve the local population's standard of living and secure their income and long term prospects.

Prospects

The vision of the Kafa BR has promoted the conservation of unique forests, highlighted cultural traditions, backed and facilitated cooperation and created new but sustainable perspectives for the region. The recent identification of people with their environmental heritage has given rise to new nature-related rituals and ceremonies and led people to identify with 'their' environment. In particular the practice of communities' protection of spiritual or sacred forest patches became widely appreciated after decades of contempt, and a special ceremonial ritual related to Thanksgiving ('Dejjo') is nowadays celebrated officially. But there are challenges that the Kafa BR is facing now and in future, such as the fight against poverty, the increasing pressure on valuable forests and the implementation of sustainable development. Nevertheless, bringing the UNESCO concept to reality offers an enormous chance to the region – and to human-kind as the unique genetic origin of Arabica coffee will be preserved and prevented from becoming extinct. The example of the Kafa BR is stimulating discussions with the aim to set up more biosphere reserves in Ethiopia: The merging of regional wildlife-friendly development, education and conservation of natural and cultural heritage may constitute a promising tool to reduce poverty in the long term.

Joining forces: The author Bender-Kaphengst in discussion with His Excellency Minister Juneydi Saddo and Prof Dr Michael Succow (© Michael Jungmeier).

NABU's mission

The Berlin-based Nature and Biodiversity Conservation Union (NABU) – the German partner of the BirdLife International global alliance of conservation organisations which is active in more than 100 countries – has been supporting the region for a number of years in close cooperation with the Ethiopian government. Founded in 1899, NABU is one of the oldest and largest nature conservation organisations in Germany. It is supported by more than 460,000 members and sponsors who are committed to the conservation of threatened habitats, flora and fauna, to climate protection and energy policy.

With reference to numerous successfully functioning biosphere reserves worldwide, NABU promotes UNESCO`s World Network of Biosphere Reserves internationally. For this reason, the organisation has held a number of high-level meetings, workshops and delegation visits in and to Ethiopia and supported the exchange of ideas and experience with German biosphere reserves. NABU's expertise was strongly backed by Prof Dr Michael Succow – an internationally renowned expert on biosphere reserves, member of the German MAB National Committee and former NABU Vice President. NABU's dedication led to a successful partnership and finally brought about a trilateral framework agreement in March 2009 between NABU, UNESCO's Cluster Office in Ethiopia and the Ethiopian Ministry of Science & Technology. Under this agreement, the three partners made the official commitment to jointly designate further biosphere reserves in Ethiopia. NABU is a member of the Ethiopian National MAB Committee. In this function, it has helped to develop, by the end of 2010, a National MAB Strategy for the country. In addition, baseline studies will be conducted for three potential new sites.

Alfonso in his farm in the Sierra Nevada de Santa Marta Biosphere Reserve, Colombia (© Lydia Thiel).

The Sierra Nevada de Santa Marta Biosphere Reserve, Colombia – The Origin of 'Coffee K.U.L.T.' ®

by Lydia Thiel & Dirk Effler

The Sierra Nevada de Santa Marta in Colombia

All over the world, we find mountains with a very special fascination for humans. Despite similarities in their appearance, each of these ecosystems is unique with its mix of endemic and immigrant life forms on one hand and its historic and current use by mankind on the other.

In 1979, UNESCO declared the Sierra Nevada de Santa Marta a biosphere reserve. The specific characteristics of this region are best described by UNESCO/MAB's own area profile (2005): 'The Sierra Nevada de Santa Marta Biosphere Reserve and National Park overlook the Caribbean coast of northern Colombia. Most of the Reserve (675,000 hectares) lies in the Sierra Nevada de Santa Marta and the remaining 56,250 hectares comprise Tayrona National Park. The area stretches from the Caribbean coast with a finely preserved coral reef, extensive beaches, several bays and inlets up to the Sierra Nevada de Santa Marta with marked relief and steep slopes. Independent of the Andean chain, it rises to a height of 5,775 meters above sea level, at a distance of only 42 kilometres from the Caribbean coast. The snowy peaks called 'tundra' are considered sacred. Vegetation ranges from sub-hygrophyte to snow levels and includes cloud forest and high barren plains. Three types of vegetation can be seen at Tayrona's National Park: forest/matorral with dry forest and humid forest. Some of them are being modified by peasants engaged in agriculture and cattle grazing, and also extraction of high-value timber, especially in the coffee belt. Of the estimated population of 211,000 (1999) some 26,500 indigenous peoples, particularly the Arhuaco, Kogui and Wiwa live in indigenous reserves, but also a considerable number live outside these areas. Ethnic groups try to develop a policy for the recovery of their ancestral lands in order to strengthen their culture and assist their traditional conservation practices. There is no management policy for the reserve as a whole and the zonation is not clear. However, scientific diagnosis and technical assessments have contributed to the elaboration of a sustainable development plan with programmes in the Sierra Nevada National Park, in agro-ecology, fish-farming and environmental health. The area is of great archaeological value particularly with sites such as the 'Ciudad Perdida' and many artifacts of Tayrona culture.' (UNESCO-MAB 2005)

Biodiversity in the Sierra Nevada de Santa Marta

As the Sierra Nevada de Santa Marta was a refuge for many species during the last glaciation, the area is characterised by an abundant diversity of plants and animal species and a high degree of endemism. About 3,000 higher plant species are found here (Tribin et al. 1999). Tapir, jaguar and puma are amongst the 120 species of mammals. The park also harbours 46 species of amphibians and reptiles. It is assumed that above 3,000 metres altitude, all amphibian and reptile species are endemic. An amazing 628 bird species have been recorded in the area of the Sierra Nevada de Santa Marta National Park alone (Rodriguez-Navarro 2007). This is approximately the number that can be found in the United States and Canada combined (The Nature Conservancy 2010).

Threats

Since the 1950s, about 85 per cent of the region's forest has been removed. Deforestation for agriculture and grazing purposes continues to be the principal threat to the Sierra Nevada. It has reduced the volume of water generated within the 35 watersheds (The Nature Conservancy 2008). During the last fifty years, the degradation of the Sierra Nevada de Santa Marta's ecosystem was severely exacerbated by inappropriate land-use practices, such as livestock breeding, illegal cultivation of drug plants (Marihuana in the 1970s and 1980s; nowadays Coca), and banana plantations in the lowlands. The air-borne campaign against illegal drug cultivation has increasingly contributed to this process. At present, only 18 per cent of the former forest area remains; two of the rivers originating in the mountains have completely run out of water. This poses a threat to both, the approximately 1.5 million people who rely on its watersheds for survival and the animal and plant species of this ecosystem (Rodriguez-Navarro 2007).

Sustainable land use to protect this unique ecosystem

Coffee (*Coffea arabica*) is grown in the 'coffee belt' at altitudes between 1,200 and 1,800 metres, in some areas even up to 2,300 metres above sea level (m.a.s.l.). In the Sierra Nevada de Santa Marta, this has been done for more than 100 years as part of extensive small-scale agro-forestry cultivation systems. Within this zone of both, high agricultural and biodiversity values, coffee-growing families contribute to the protection of some of the remaining original forests by using them to serve as shading for their coffee plants. Land use and biodiversity protection can go hand in hand as shown by the example of 'Cuchilla de San Lorenzo'. The area 'Cuchilla de San Lorenzo' which extends from the village of Minca (600 m.a.s.l.) across the area of coffee plantations to the border of the Sierra Nevada de Santa Marta National Park, was declared an 'Important Bird Area (IBA)' by BirdLife International (2009) and the Alexander von Humboldt Biological Resources Research Institute which has coordinated the IBA programme in Colombia from the outset in 2001. The site has also been identified as 'Alliance for Zero Extinction Site' as it contains several endangered bird species with a limited distribution range, such as the Santa Marta parakeet or bush-tyrant (AZE 2010).

ALPEC strives for conservation by sustainable use

The Colombian foundation 'Alianza para Ecosistemas Criticos' (ALPEC) is running projects which aim at the protection of the ecosystem 'Sierra Nevada de Santa Marta' and its biodiversity. In order to guarantee the compliance with its conservation premises, ALPEC has designed and implemented the certification system 'Critical Ecosystem Alliance' (CEA). The system regards itself as a process for enhancing agricultural practices for the benefit of both, natural ecosystems and producers. The signet of certification acknowledges sustainable production in the sense of protection of wild flora and fauna. The criteria of certification and therefore agricultural production and processing have been developed in close cooperation with the producers. ALPEC is also aiming at the creation of ecological corridors, as well as sensitisation and persuasion of local communities and producers. ALPEC was initiated as a project more than ten years ago by Dr Ralph Strewe, then lecturer at the University of Santa Marta, and has since adopted its current structure as an NGO.

'Coffee K.U.L.T®' and the idea of partnership

Coffee-growing is one of the most promising and sustainable economic activities in the Sierra Nevada de Santa Marta. The coffee belt – the zone in which *Coffea arabica* can be grown – comprises an area of approx. 168,000 hectares. The environmental qualities of the coffee belt of the Sierra Nevada de Santa Marta match perfectly the land use requirements of *Coffea arabica*, especially such factors as altitude, precipitation, soils and temperature. Shade is provided by the native rain forest trees (Effler 1992). At the same time, these trees provide the necessary habitat for endemic and other protected species as well as for migratory birds. 'Partnerschaftsprodukte e. V.' (partnership products) is a German association which promotes and distributes

School children participate in the afforestation of degraded areas: Lydia Thiel and teachers posing for 'plants for the planet' (© Lydia Thiel).

products from UNESCO Biosphere Reserves in order to support the protection of rain forests. These products are intended to contribute directly to the protection of fragile ecosystems and biodiversity as well as enhancing the sustainable livelihood of small-scale farmers. 'Partnerschaftsprodukte e.V.' is engaged in a partnership with ALPEC, and provides support to their projects in Colombia. This association is also involved in marketing products of the Sierra Nevada de Santa Marta in German-speaking parts of Europe. The association has intensive direct contacts with Colombian, German and international organisations as well as local producers. Current projects aim at supporting the sustainable use of the rainforest within the biosphere reserve, the afforestation of degraded areas, the reintroduction of indigenous tree species to degraded forests (in close cooperation with the organisation 'Plants for the Planet'), and the marketing of the 'coffee K.U.L.T.®' brand (a registered trademark of Partnerschaftsprodukte e.V.) linked with the message of rainforest conservation both, in Europe and in Colombia. Major activities include raising awareness in both countries, fostering communication and cooperation between schools in Colombia and Germany, supporting direct contacts and partnerships between coffee growers and (small-scale) coffee roasters in Europe. Future projects are intended to generate scientific exchange and to trigger more direct partnerships between stakeholders in Colombia and Germany.

Coffee-growing families participating in the project (currently 14) pledge to dedicate protected areas within their farms (fincas), to enrich existing forests with native tree species, to protect rivers sources and water courses, as well as to abandon any chemical products. A variety of training schemes are made available. Project experience has shown that ecological production methods and direct marketing can offer better income opportunities for small-scale farmers. This, in combination with greater economic independence and self-confidence of small-scale farmers in a difficult socioeconomic environment can lead to a more sustainable use and the protection of rainforest resources. 'Partnerschaftsprodukte e.V.' imports coffee directly from the Sierra Nevada de Santa Marta Biosphere Reserve with high expectations in terms of quality and ecological, social and economic standards as prevalent in Europe. By buying 'coffee K.U.L.T®', consumers develop a heightened awareness of the protection of tropical rain forests, take responsibility and contribute to the protection of biodiversity.

Future prospects

In the near future, 'Partnerschaftsprodukte e.V.' and ALPEC intend to identify and support partner schools in Germany and Colombia and foster contacts between biosphere reserves in Europe and the Sierra Nevada de Santa Marta BR. Another important cooperation has just started between the Technical University of Munich and scientific partners in Santa Marta. Further efforts in increased marketing of 'coffee K.U.L.T.®' are intended to spread the ideas of the protection of rain forests, biodiversity and climate within German-speaking parts of Europe.

Conclusions

As Europeans, we can actively influence and contribute to the protection of 'rain forest' habitats and to the conservation and development of the livelihoods and culture of people living in these regions. We can do this, for example, by consuming 'coffee K.U.L.T.®'. It is imported directly from the local 'finca', carefully roasted in Germany in family-run businesses and marketed through organisations which support the ideas of 'Partnerschaftsprodukte e.V.' (2010).

Package of Coffee K.U.L.T.® which is sold in German-speaking countries (© Lydia Thiel).

References

AZE – Alliance for Zero Extinction (2010). Information on the Parque Nacional Natural Sierra Nevada de Santa Marta at http://www.zeroextinction.org/sitedata.cfm?siteid=391 (accessed on 12 October 2010).

BirdLife International (2009). Factsheet on the Important Bird Area 'Cuchilla de San Lorenzo', Colombia; downloaded from the data zone at http://www.birdlife.org/datazone/sites/index.html?action=SitHTMDetails.asp&sid=14418&m=0 (accessed on 12 October 2010).

Effler, D (1992). Landnutzungsplanung in ASOMAYO (Kolumbien). Landeignungsbewertung für kleinbäuerliche Betriebssysteme im andinen Raum nach der FAO – Methodik. Berlin.

Partnerschaftsprodukte e.V.' (2010). Internet portal at http://www.partnerschaftsprodukte.de

Rodriguez-Navarro, G. E. (2007). Conflict, traditional knowledge and conservation: Sierra Nevada de Santa Marta case study. Download at: http://peaceparks2007.whsites.net/Papers/Rodriguez-Navarro_Colombia.pdf

The Nature Conservancy (2008). Parks in peril. Sierra Nevada de Santa Marta National Natural Park. Information at http://www.parksinperil.org/wherewework/southamerica/colombia/protectedarea/sierra.html (accessed on 20 October 2010)

The Nature Conservancy (2010). Colombia. Places we protect. Sierra Nevada de Santa Marta. Information at http://www.nature.org/wherewework/southamerica/colombia/work/art5303.html accessed on 21 October 2010

Tribin, M., Rodriguez-Navarro, G. & Valderrama, M. (1999). The Biosphere Reserve of the Sierra Nevada de Santa Marta: A pioneer experience of a shared and coordinated management of a bioregion. Working Papers N°30. South-South-Cooperation Programme on Environmentally Sound Socio-Economic Development in the Humid Tropics

UNESCO-MAB (2005). Biosphere Reserves Directory, information on Sierra Nevada de Santa Marta at http://www.unesco.org/mabdb/br/brdir/directory/biores.asp?mode=all&code=COL+03 (accessed on 20 October 2010).

The municipal office of St. Gerold in Großes Walsertal BR was constructed as a 'passive house' (© BR Management).

Increasing Energy Efficiency – the Case Study of Großes Walsertal Biosphere Reserve

by Sigrun Lange

In 2007, the Intergovernmental Panel on Climate Change (IPCC) released its fourth assessment report, demonstrating that global average air and ocean temperatures are rising with the effect of increasing runoff and earlier spring peak discharge in many glacier-fed and snow-fed rivers, increasing ground instability in permafrost regions, and rock avalanches in mountain regions (just to mention a few of the impacts observed). Computer models such as the one used in the ALARM project (Assessing LArge scale environmental Risk for biodiversity with tested Methods) are predicting that, based on the assumption that the mean temperatures will increase by four degrees Celsius by the end of this century, approx. 20 per cent of the species in Europe may lose about 80 per cent of their current distribution areas (Thuiller et al. 2005). By contrast, at high altitudes, overall bio-diversity, of vascular plants in particular, will increase. This is due to the fact that the combination of less snow and higher temperatures will improve the living conditions for plant life. However, the species adapted to the uppermost reaches will lose most of their habitats (Grabherr, Gottfried & Pauli 2010).

The international community is very concerned. In the context of the 'United Nations Framework Convention on Climate Change' (UNFCCC), Article 193 recommends[1] joint debates on what can be done to reduce global warming. The next meeting will be held in December 2010 in Cancun, Mexico. Likewise, UNESCO's MAB Programme identified climate change as one of the '*most serious and globally significant challenges to society and ecosystems around the world today. The role of biosphere reserves (BRs) is essential to rapidly seek and test solutions to the challenges of climate change as well as monitor the changes as part of a global network*' (UNESCO MAB 2008, p.6).

Großes Walsertal Biosphere Reserve

The 'Große Walsertal' is a sparsely populated mountain valley in the Federated State of Vorarlberg (in western Austria). It is inhabited by approx. 3,400 people who live in six communities: Fontanella-Faschina, St. Gerold, Raggal-Marul, Sonntag-Buchboden, Thüringerberg and Blons. For a long time, cattle raising was the main economic activity. Nowadays, farmers focus on organic farming and an ecologically sustainable use of the mountain forests. Meanwhile, tourism plays a major role (hiking in summer, small ski lifts for winter tourism). In 2000, the mountain valley was designated a UNESCO biosphere reserve. From the very beginning, the planning process was conducted in a participatory approach. The inhabitants were invited to propose a mission statement for the future development of the valley.

Efforts towards energy efficiency

The six communities of Großes Walsertal BR take their function – to act as 'living laboratories' for testing sustainable solutions – seriously. In February 2010, the BR was awarded the 'European Energy Award®' in silver. The Award is the highest recognition in Europe awarded to energy efficient communities. It is a clear recognition of the continued efforts made by the communities in this mountain valley on the way to achieving their goal of energy self-sufficiency, and becoming an export region for renewable energy by 2030. Back in 2009, in the course of the 'e-Regio' project – financed by the Austrian Climate Fund – a comprehensive package of measures was indeed developed. The future energy strategy will be based on a balanced mix of biomass, increased energy efficiency, hydroelectric power and soft mobility.

Biomass (in form of wood chips) will be the main material used for heating. Timber grows in the mountain forests of the region; the transport routes can therefore be kept short. In 2003, in Fontanella-Faschina, a biomass heat supply station was built to heat all the hotels present in the small ski resort. Every year, 200,000 litres of fuel oil are saved this way. In addition, tourists can book a guided tour of the power plant to obtain some insights into renewable energies. In 2006, another biomass power plant was opened in St. Gerold. In the same

---

[1] Member states of the UNFCCC (accessed 24 October 2010): http://unfccc.int/essential_background/convention/status_of_ratification/items/2631.php

year, a storage shed for wood chips was built in Raggal. Now, the material is available on the site where it is needed. All six communities have local heat networks, and therefore it is already possible now to supply enough biomass to cover 80 per cent of the heat demand for communal buildings and 60 per cent of the heat demand for private buildings. The intention is that by 2020, the use of biomass from the region will create a valley free of fuel oil. The 'e-Regio' Working Group on Biomass recommended building a drying plant for wood chips and adapting long-term delivery contracts to the actual demand. In order to increase energy efficiency, three per cent of the housing stock is to be improved in terms of energy consumption by means of thermal insulation. New buildings are constructed in the most energy-efficient way; for example, St. Gerold's Municipal Office was constructed as a 'passive house'.

Sufficient water resources and ideal pressure heads are the best preconditions for using hydroelectric power. Several small privately owned hydro-power stations are already in operation in the Große Walsertal. They produce more than ten million kilowatt-hours, and cover approx. 71 per cent of the energy consumption in the BR (October 2009)[2]. Photovoltaic panels add to the power supply in the BR which is already covering its entire demand for electricity from renewable energy. In 2003, one of the largest flexible photovoltaic systems worldwide was constructed in Blons. Electric motors automatically shift the total of 21 panels from facing east to facing west, thus following the course of the sun. The automatic system is controlled by light sensors. In case one of the panels is shaded by a neighbouring panel, it is automatically lifted towards the sun. In case of heavy snowfall, the panel will be brought into a near vertical position (Krampitz 2003). The 'e-Regio' Working Group on Hydro-power recommended optimising the existing plants and testing the potential for introducing drinking-water power stations. For the latter, no new designs will have to be built. The water containers and conduits already exist; only the surplus water would be used for generating electricity.

Achieving mobility in remote mountain regions

Generally, facilitating mobility without increasing individual motor car traffic is one of the greatest challenges in remote mountain valleys. For several years, the communities in the Großes Walsertal BR have tried to improve the public transport system. The BR can be reached by a combination of trains and local buses. During the hiking season in summer and autumn, additional buses take visitors to the starting points of hiking routes in some of the Alps. In recognition of this achievement, the Großes Walsertal BR was included in the list of 17 so called 'hikers villages' (Bergsteigerdörfer) in Austria[3]. Started by the Austrian Alpine Club (OEAV), this initiative tries to promote villages in the Alps with outstanding landscapes and wildlife, well-preserved traditions, and green tourism facilities, including

---

Information on the European Energy Award

The European Energy Award® (eea®) is a suitable instrument for steering and controlling communal energy policy in order to review systematically all energy-related activities. The award allows municipalities to identify strengths, weaknesses and potentials for improvement and, above all, implement energy efficient measures in an effective manner. The standardised assessment sets a benchmark for all eea® communities. In a step-by-step process, communities improve their performance in terms of their energy-related activities by:

• Reviewing energy-related activities;

• Visualising strengths, weaknesses and potentials for improvement;

• Defining goals for the local energy policy and defining decision-making criteria;

• Developing an energy policy work programme comprising concrete long-term and short-term projects;

• Step-by-step implementation of the work programme;

• Continuous assessment of the results.

The entire process is carried out by the energy team, formed by representatives from the communities' administration and politicians, assisted by an external eea® advisor who is an expert in the field of energy.

Currently, 755 communities in seven European countries participate in the European Energy Award®. Information is available at: http://www.european-energy-award.org.

---

public transport to and within destinations. The 'e-Regio' Working Group on Mobility suggested to further improve the public transport system and to restructure it in the long-term. A mobility centre is to be established in the BR for coordinating new facilities, such as car-sharing, using electric cars. One proposal is to complement public transport by offering electric bikes to tourists. One e-bike can already be rented and tried out at the BR's management office.

During the hiking season, special buses take visitors to the starting points of hiking routes (© BR Management).

The Großes Walsertal BR can rightly be considered a mountain region that makes an exemplary contribution towards mitigating global warming. At this stage, it is already possible for visitors to learn from local experience in guided tours. In future, it is planned to operate an 'energy house' in cooperation with the local trade and commercial businesses. This will be the stage in which information on energy efficiency and sufficiency can be exchanged in order to stimulate the willingness of the greater public to modify our behaviour in the interest of a more sustainable future.

Another exemplary alpine biosphere reserve

However, Großes Walsertal BR is not the only biosphere reserve in the Alps engaged in increasing energy efficiency. The Swiss Entlebuch BR, established in 2001, is also aiming high. About 17,000 inhabitants live in the mountain valley between Bern and Luzern. The 'energy forum', one of the working groups in the BR, has set itself several goals to be reached by 2020. One of these objectives is to increase energy efficiency in municipal, commercial and private buildings, achieving a CO_2-neutral heat energy balance, and covering 20 per cent of the electricity demand from local renewable sources (15% from a hydroelectric power plant, 5% from wind energy). In 2003, a wind energy concept was proposed for the UNESCO site. Subsequently, in October 2005, the first wind power station was put into operation above the village of Entlebuch. In addition, the potential of drinking-water power stations was analysed. As a result, two projects have been implemented in Sörenberg and Schüpfheim (Entlebuch BR 2010).

In conclusion, it is to be hoped that BRs worldwide will take their role seriously, acting as pilot regions for trend-setting solutions related to global warming and a sustainable lifestyle. The activities in the Großes Walsertal and Entlebuch BRs may stimulate further projects in the mountain regions of the world thus helping to implement on a local level the decisions taken by the international community as for example this year in Cancun.

References

Entlebuch Biosphere Reserve (2010). Information of the 'Energieforum' at the official web site of the Entlebuch BR at: http://www.biosphaere.ch/de.cfm/company/forums/offer-GesellschaftUBE-Foren-323406.html (accessed on 28 October 2010).

Grabherr, G., Gottfried, M. & Pauli, H. (2010). Climate Change Impacts in Alpine Environments. In: Geography Compass 4/8 (2010): 1133–1153.

IPCC (ed.) (2007). Climate Change 2007: Synthesis Report. Available at: http://www.ipcc.ch/publications_and_data/publications_ipcc_fourth_assessment_report_synthesis_report.htm (accessed on 24 October 2010).

Krampitz, I. (2003). 420-kW-Nachführanlage in Österreich eingeweiht. In: Photon, das Solarstrom-Magazin Online. 24.08.2003. Available at: http://www.photon.de/news/news_panorama_03-09_nachfuersystem.htm (accessed on 27 October 2010).

Thuiller, W., Lavorel, S., Araujo, M.B., Sykes, M.T. & Prentice, I.C. (2005). Climate change threats to plant diversity in Europe. In: Proceedings of the National Academy of Sciences of the United States of America 102 (23): 8245–8250.

UNESCO MAB (2008). Madrid Action Plan for Biosphere Reserves (2008 – 2013).

Worldwide Case Studies

Model Regions for Participatory Approaches, Global Networking and Effective Evaluation

Learning from experiences

Reforestation measures in Sierra Gorda BR, Mexico
(© Roberto Pedraza, www.sierragordasilvestre.net).

Education and Training for Conservation: the Case of the Sierra Gorda Biosphere Reserve, Mexico

by Roberto Pedraza

In 1989, a social movement supported by multiple stakeholders began lobbying for the conservation of the Sierra Gorda region in the state of Querétaro, Mexico, with the aim of designating the mountainous region a protected area. The efforts paid off: In 1997, the Sierra Gorda Biosphere Reserve was created by presidential decree. Four years later it was internationally designated by UNESCO. In one respect the reserve is unique: The movement, organised in the Sierra Gorda Ecological Group (GESGIAP), has become the operational arm of the managing authority which has turned the area into the only model of a biosphere reserve co-managed by a social initiative and the National Commission of Natural Protected Areas (CONANP). This structure has stimulated unprecedented social participation. The Sierra Gorda Biosphere Reserve turned out to be the protected area in Mexico with the greatest number of people involved in conservation activities.

Engagement of GESGIAP in the development of the BR
The GESGIAP began as a small and humble local initiative founded by citizens concerned about the rapid deterioration of an area known for its exceptional biological richness. At first, the group was mainly engaged in environmental education and reforestation. Relationships with the local society were built, a process that still continues and has been the basis of all undertakings. Since then, the movement has grown into a conservation project with significant outreach to national and international levels, breaking new ground in many fields of action and receiving important recognition from its grassroots. In 1999, the GESGIAP developed a management plan for the Sierra Gorda BR. In cooperation with the Secretariat of Environment and Natural Resources (SEMARNAT), it managed to obtain approval from the Global Environment Facility (GEF) for the implementation of the project 'Biodiversity Conservation in the Sierra Gorda Biosphere Reserve'. The project was administered by the delegation of the UN Programme for Development in Mexico, operated by the National Commission of Protected Natural Areas, and executed by the Sierra Gorda Ecological Group. This marked a turning point for the former grassroots organisation. It managed to acquire financial resources from

GEF and matching funds amounting to a total of $48 million by the time the project ended. It was the first time in the history of conservation in Mexico that a NGO obtained approval for a full-scale project by the GEF and greatly exceeded the requirements for matching funds. With these funds, the organisation was able to accomplish all 167 lines of action foreseen in the management plan. The implementation of the project was possible thanks to a strong network of partners and allies at national and international levels, ranging from the three levels of government which developed criteria for sustainable development, to a variety of foundations and organisations. The project strived for complex and diverse objectives, such as strengthening local capacities by education and training, developing a curriculum for the Sierra Gorda Earth Centre, establishing a geographical information and monitoring system, developing a payment system for environmental services for the benefit of landowners in the area, restoring micro-watersheds, reforesting cutover areas, enhancing sustainable cattle breeding, diversifying products and services in the region, carrying out inventories of carbon dioxide stored in local ecosystems, selling CO_2 certificates under a voluntary mechanism, and establishing private nature reserves. In particular, capacity building was a substantial line of the project. Educational activities and training were offered by the Sierra Gorda Earth Centre for different stakeholders, such as producers, housewives, and personnel from other protected areas or government agencies.

Training and education for conservation
Within the scope of the 'Biodiversity Conservation in the Sierra Gorda Biosphere Reserve' project, and its Environmental Education Programme, the GESGIAP has implemented a variety of educational activities ranging from the design of leaflets or posters to radio broadcasts, murals or blankets with messages. The material reached about 18,000 students in 172 schools of basic-level education and 110 communities in the five municipalities involved in the Sierra Gorda BR. Already before, in 1989, a significant number of adults, including producers, housewives, merchants, local authorities and landowners, was trained in environmental topics. Hence, as a result, more and

more stake-holders gradually adopted new habits while discarding attitudes and practices harmful to nature. All these actions have had a major impact and contributed to the conservation of natural resources within this green jewel. Currently, the environmental school education programme is being extended from pupils to local teachers. 600 teachers participate voluntarily in various forms of training. They take an active part in the organisation of approx. 50 Earth Festivals annually that bring people together and strengthen the link between communities and schools. GESGIAP continues to support the curriculum, the educational materials as well as radio promotion. It is of capital importance, therefore, to arouse interest and enthusiasm, and feel comfortable within a network of supporters and friends of conservation.

The Earth Centre is the main platform for environmental education. It offers courses, workshops and diplomas, both on the ground and on-line, with the support of the Technological University of Querétaro operating though its virtual campus platform. Since February 2009, the online diploma course 'Learning and Teaching for a Sustainable Future' has been made available. Based on the materials from the UNESCO Decade of Environmental Education, it was expanded on the basis of lessons learned from the Sierra Gorda project. To date, 1,500 people have been trained in a total of 35 courses and workshops, with participation by 29 different protected areas and staff of CONANP. The courses have been generously supported by institutions like CONANP, SEP, CECADESU, UAQ and foundations such as the National Monte de Piedad, the U.S. Fish and Wildlife Service, and the Mitsubishi Foundation for the Americas.

Sustainable use of resources: refuse separation and recycling

In 1992, a programme on the sustainable use of resources was launched. The local population was pointed to the problem of far too much refuse. The urgent need to separate the solid waste in order to reduce the amount of refuse in landfill sites was explained. Concrete actions such as community awareness meetings, cinema shows, clean-up and waste separation campaigns were carried out. Currently, activities are occurring in 115 collectives within the biosphere reserve. 110 local and two regional refuse collection centres have been installed. Each collective has a committee which is linked with the committees of the other communities. It is the only rural network of this type operating in Mexico. Moreover, in future, it is intended to link this network with other institutions such as the National Institute for Adult Education, the National Commission on Educational Development, Early Childhood Education, the Opportunities Programme, and the Health Sector. A major achievement of the refuse project is the separation of 500 metric tonnes of solid waste per year that do not end up in the landfill sites of the Sierra Gorda region. An enhanced management of landfills and the installation of water treatment plants by the State Government have contributed to the improvement of refuse disposal in the Sierra Gorda.

Rare animals such as the margay still roam the Sierra Gorda BR (© Roberto Pedraza, www.sierragordasilvestre.net).

Characteristics of the Sierra Gorda BR

The Sierra Gorda BR was created by presidential decree on 19 May, 1997, with the purpose of protecting the exceptional richness of species and ecosystems of the mountainous area. It is located in the north of the Mexican State of Querétaro, and covers an area of 383,567 hectares, representing 32 per cent of the state´s territory. It comprises eleven core zones totalling 24,803 hectares (6.5%), and a buffer zone that covers an area of 358,764 hectares (93.5%). Almost 95,000 inhabitants live in the area, and five municipalities are involved (Jalpan de Serra, Arroyo Seco, Landa de Matamoros, Pinal de Amoles and Peñamiller).

Owing to its geographical position at the convergence of the Nearctic and Neotropic Bioregions and its topography with elevations ranging from 300 meters above sea level in the Santa Maria River Canyon up to 3,100 meters above sea level in the Cerro de La Pingüica, the Sierra Gorda region harbours an abundant diversity of plant and animal species. It is the best preserved and the most diverse area in Querétaro. The vegetation includes approx. 2,300 species of vascular plants. In view of the fact that a number of highly endangered species live in the area, the conservation of Sierra Gorda ecosystems is essential. The Reserve also serves as a refuge for migratory species. The vertebrate faunal diversity is composed of 111 species of mammals, 334 species of birds, 97 of reptiles, 34 of amphibians, and 27 species of fish six of which are considered endangered.

Currently, the community refuse project is in the process of being expanded to local City Councils. They can benefit from an already existing network of customers who purchase the recyclable materials (glass, plastic, cardboard, aluminium and iron) from the Sierra Gorda region. These organisations include CODSI, a regional cooperative which covers the five municipalities of the BR, and the Plastilanda Cooperative which consists mainly of women.

The provision of economic alternatives for traditional practices with high environmental impacts and little earnings (e.g. logging of trees of low commercial value) has been a constant effort since the designation of the Biosphere Reserve. Destructive practices are gradually replaced by more eco-friendly activities. The BR has trained, equipped, and given follow-up support to different local groups, which has opened up a variety of options for generating income. For example, a label for Sierra Gorda products has been created. It covers all products manufactured by rural-community micro-enterprises, such as embroidery with wildlife motifs, ceramics, herbal products, honey, organic food and products from five pilot farms with sustainable cattle ranching. Currently, 363 people benefit directly from micro-enterprises (consisting of one third women), and 1,452 benefit indirectly. In places of outstanding beauty, a net-work of nine eco-lodges has been established. They are owned and operated by the communities themselves. Sierra Gorda Ecotours functions as a tour operator which brings clients to the lodges. Regional products are included in tourist packages in order to create attractive offers for all types of tourists and students. Consolidating the network has been a long-term effort to retrain local foresters or farmers to become tour operators. Training has been given by staff of Sierra Gorda Earth Centre and to some extent by experts from outside the area.

Conclusions

Thanks to a bottom-up development and the strong involvement of local land owners and user groups, the Sierra Gorda BR can be considered an effectively managed protected area. The model of co-management by private organisations that are strongly rooted in the area, in cooperation with the Federal Government has proved successful and, in fact, has turned out to be the only way to ensure conservation in a country with insufficient budgets assigned to protected areas. In Sierra Gorda, a model has been established with many repeatable experiences and lessons learned some of which are adaptable to local conditions in other areas of Mexico and especially in Latin America. Twenty-one years of experience in environmental education have left their mark on training in Sierra Gorda's Earth Centre, and this can serve as a basis in Latin America for disseminating practices and knowledge regarding sustainable use.

Children of Sierra Gorda (© M. Bertzk

Key Factors for a Successful Implementation of the Biosphere Reserve Concept – the Example of the Mexican Sierra Gorda

by Monika Bertzky

In the 'Big Mountains' of Querétaro, the Mexican 'Sierra Gorda', a Biosphere Reserve was established in 1997 which has since evolved into a model region that reflects the vision of the World Network of Biosphere Reserves as outlined in the Madrid Action Plan (UNESCO 2008). But how did it get there? A series of 30 interviews with representatives from the national government, the BR management, local NGOs and local community members has shed some light on factors that have facilitated this development[1] (for more detail see Bertzky 2009). While it may seem common sense to realise that the combination of these factors will favour a successful implementation of the BR concept, finding them in practice is not common and cannot be taken for granted.

Key factors

The governance regime of the Sierra Gorda BR (SGBR) is characterised by co-management between a social initiative and the national government. Thanks to the long-term commitment of all key-actors involved, this co-management regime

has gained in strength and stability. The social movement started more than 20 years ago, and the key actors in this initiative have remained the same. They have established a close relationship with the people of the region. Having grown up in the Sierra Gorda themselves, the people concerned have developed an intrinsic feeling of responsibility for the place that is their home. A true understanding of the people's issues, worries and concerns has helped them to identify socially viable, more sustainable opportunities for alternative income, which they now help to put into practice. Where such favourable conditions are matched by strong leadership, as in the case of the SGBR, a lot can be achieved. One interviewee describes the phenomenon as follows:

'*The interest [that key actors of the SGBR have] in conservation is what makes a lot of things move, what detects the spaces, the capacities, not only on a local level, but even on an international level. [...] Without doubt those motivators are essential, there are many in many places of the world, but well, we have the fortune that one of them is here.*'

Another important success factor has been the concerted action at various governance levels. The SGBR is very

[1] This study was conducted in 2006 as part of a PhD thesis within the Governance of Biodiversity (GoBi) Project with financial support from the Robert Bosch Stiftung.

actively working at the local level, implementing conservation and awareness-raising activities, providing capacity building and supporting alternative options for making a living, reaching out to several thousands of people living in scattered communities across the mountains. However, activities go beyond the local level and include regional and national government offices; they also involve international organisations, universities and funding agencies. This has proved helpful in gaining regional and national support, as well as international recognition for the achievements on the ground. The feedback from external agents has helped local people and local and regional decision-makers to develop more and more personal pride in being part of this movement. It also spurs the willingness to engage evermore actively in conservation activities. The following statement emphasises his observation:

'It is not only students, […], everybody is involved in environmental clean-up until up to the authorities; something very important, the change of the attitude within authorities who now also have their own actions to the benefit of the environment, to the benefit of sustainability. It is a fact that now the public works are much more focussed on sustainability.'

Appropriate and sustainable resourcing is a success factor whose importance should not be overlooked, and in the case of SGBR, significant support has also been secured through successful communication and cooperation with international partners. Without the SGBR's large number of staff and equipment, of which vehicles are central, it would be simply impossible to reach out to so many inhabitants in an area of poor infrastructure where settlements are widely dispersed and driving to various places is arduous and time-consuming. If such a level of resources can be maintained sustainably, environmental education and awareness-raising can be achieved with stable or increasing efforts over a long period of time. This is essential for achieving an attitude shift towards more sustainable thinking, and eventually also living. A civil servant from the national government described the success of these efforts within the BR by comparing it with a neighbouring state:

'I don't know whether you passed by San Luis Potosí, […], I mean, you would notice that the border of the state of San Luis Potosí is not only a political border, it is also an ideological frontier, an ethnic frontier.'

However, other neighbouring states, inspired by the large number of conservation, awareness-raising, education and capacity-building activities taking place in the SGBR, have become increasingly interested in securing designation of a BR themselves. Following successful application, the 'Sierra Gorda de Guanajuato BR' was recognised as a national-level Biosphere Reserve in February 2007.

Common sense – not to be taken for granted

It may not come as a surprise for you to hear that the combination of time, commitment, stability, knowledge and understanding of local circumstances, strong leadership, strategic action on different governance levels, and appropriate and sustainable resources for conservation actions, environmental education and awareness-raising facilitates the successful implementation of the BR concept. However, there are many BRs in the world that lack one, some, or all too many of these key factors (Stoll-Kleemann and Welp 2008). Insufficient resources restrict action, and the permanent involvement of key players all too often depends on the outcomes of political elections. Repeated changes in management positions can entail changing practices and priorities. This can destabilise good and trusting relationships with local people. Moreover, patience is a rare commodity in these fast-moving times where pressures on resources often require quick conservation action. In addition to the challenge of implementation alone, such actions are expected to be monitored in order to make it possible to prove success in statistical terms. But getting conservation going is time-consuming and it takes even longer before the fruits of such efforts become evident. In the past, the SGBR has also had its fair share of difficult times. The availability of resources could not always be relied on, and relationships with decision-makers were not always stable. The combination of time and the other factors noted above have helped to overcome these and other hurdles and made it possible eventually to turn the SGBR into a model region for the implementation of the BR concept. Local people have reported more sightings of wildlife (Bertzky 2009), deforestation has been reduced (de la Llata Gómez 2006), and camera traps keep providing evidence for the presence of iconic species such as jaguar in the SGBR. A civil servant in the national government summarises the success of the BR in one sentence: 'The state of conservation of the Sierra Gorda is much better than I would have been able to imagine.'

References

Bertzky, M. (2009). Mind the gap: Information gaps and bridging options in assessing in-situ conservation achievements. Dissertation. Ernst-Moritz-Arndt University Greifswald, Germany. Available from http://ub-ed.ub.uni-greifswald.de/opus/volltexte/2009/575/.

De la Llata Gómez, R., Bayona Celis, A., Rivera Sánchez, E., Guadalupe Valtierra, J., Martínez-Reséndiz, W.E. & Montoya Martínez, A. (2006). Vegetación, Uso del Suelo y Unidades de Paisaje en la Sierra Gorda Queretana. TOMO XI, Reporte Técnico. Centro Queretano de Recursos Naturales, Consejo de Ciencia y Tecnología del Estado de Querétaro, Querétaro, Mexico, 54 pp.

Stoll-Kleemann, S. and Welp, M. (2008). Participatory and Integrated Management of Biosphere Reserves. Lessons from Case Studies and a Global Survey. GAIA, 17(S1): 161–168.

UNESCO (2008). Madrid Action Plan for Biosphere Reserves (2008–2013). UNESCO, MAB Programme. Paris, France.

Postglacial cirque in Krkonose/Karkonosze Transboundary Biosphere Reserve (© KRNAP).

Biodiversity Conservation in the Transboundary Biosphere Reserve of Krkonose/Karkonosze

by Jiri Flousek & Jakub Kaspar

The Krkonose (Giant) Mountains are a mountain range in NE Bohemia and SW Poland split in two by the Czech-Polish border. They are famous for extremely high terrain, geological and species diversity in four altitudinal belts ranging from submontane to alpine. Even though not very high, the mountains appear as a sole ecological island of arctic and alpine ecosystems whose counterparts are located many hundreds of kilometres to the south in the Alps, and to the north and northwest in Scandinavia and the British Isles. The extraordinary natural values of Krkonose derive from their geographical location in the centre of Europe, from a remarkable geomorphology and a harsh climate. The mountains have played the role of an extremely important 'crossroads' which acted as a link between the northern tundra, pushed further and further south by continental glaciers, and the alpine and subalpine ecosystems that expanded from the Alps northwards. More than 1,300 taxa of vascular plants, including many endemics and glacial relics, have been identified in the most valuable habitats: alpine tundra, subarctic peatbogs and glacial corries, flower-rich mountain meadows, dwarf pine stands, mountain spruce forests and remnants of autochthonous mixed beech-spruce forests. Plant diversity is of vital importance to the rich fauna. The proportion of glacial relics among invertebrates is high, especially compared to the nearest mountain ranges. On the other hand, the level of endemism is very low (three taxa only). About 270 vertebrates have been registered recently: among them more than 150 breeding bird species and about 60 mammalian species.

Large-scale forest destruction caused by air pollution was the main problem at the end of the 20th century. Currently enormous pressure on the natural environment is caused by tourism and recreation. On the Czech side of the mountains, approximately five to six million visitors are recorded annually. Another two million visit the Polish side. There are important summer and winter recreation resorts within the transition zone, and their development entails increasing pressure on economic use of the area. As a result, natural communities are widely influenced by a complex assortment of negative impacts such as recreation and accompanying activities (chairlifts, downhill ski courses, chalets, refuse collection, eutrophication, etc.).

Long-term transboundary cooperation on conservation

The history of Czech-Polish cooperation on nature conservation dates back to 1925, when scientists from both countries signed the so-called 'Krakow Protocol' which pioneered the establishment of the Karkonosze National Park (PL) in 1959, followed by the Krkonose National Park (CZ) in 1963. The need of transboundary cooperation was appreciated by both park authorities and their respective governments. By signing an agreement in 1988 they committed themselves to take measures against environmental damage on both sides. After the political changes of the early 1990s and the opening of the borders, bilateral contacts were extended from the top management level to the rest of the staff (at first mostly biologists, rangers and foresters).

In 1992, the whole mountain range was declared the UNESCO's Transboundary Biosphere Reserve Krkonose/Karkonosze, as one of the first bilateral biosphere reserves worldwide. It covers the area of the two national parks, with an additional transition zone on the Czech side. Subsequently, the national parks staff was assigned conflicting tasks: As staff of the national parks they were mainly responsible for nature conservation (which often means restricting human development), whereas as representatives of the Transboundary Biosphere Reserve they had to stimulate solutions for sustainable development. As it was quite difficult to draw a line between the two remits, it was decided to separate the objectives by creating different bodies. In 1996, an agreement was signed to create a Czech-Polish BR Board for all transboundary issues with respect to sustainable development and environmental policies.

Since 1997, the so-called 'Bilateral Council of the BR' (BCBR) is to meet once a year. Each country delegates representatives to these meetings from national parks, local councils, state administrations, private sector, NGOs and from five Czech-

Polish working groups which were established in the fields of
* nature conservation (conservation on ecosystem/ community/
 species level, inventories, monitoring and research, data management, GIS, etc.),
* forest management (forest plans, etc.),
* recreation & tourism (tourist management, visitors' survey etc.),
* public relations (ecological education, culture, publication activities etc.) and
* socio-economic development (landscape planning, agriculture, industry, etc.).

Czech and Polish secretariats were established as coordinating bodies of the BCBR (one person per country, financed from different grants and foundations). The coordinators represent the transboundary biosphere reserve vis-à-vis the local population but in legal terms the biosphere reserve is managed by the staff of the two national parks. They are assisted by the working groups mentioned above, who meet when required. However, both the Czech and Polish secretariats have had no staff recently, so that part of their duties is again performed by staff from the national parks. To be frank, the BR Council too ceased its activities (as a result of reduction in the work carried out by the two secretariats. One reason for this was the intention to strengthen communication and cooperation between the two NPs in many other fields which had been quite demanding in terms of human resources). This situation is to be remedied now, and we want to tackle this task within the next few months. Another international step forward is connected with the conservation of wetlands. Subarctic peatbogs on the ridge of the mountains along the Czech-Polish border have been listed as bilateral Ramsar Convention sites since 2009.

Great work has been done in cooperation between the national parks and local councils when jointly preparing the 'Vision for Krkonose 2050'. Initiated by the Czech Board of the NP, it was from the outset prepared with active Polish participation. The vision has been agreed by the Board now and mentioned in various planning documents published by the NPs and local councils (e.g. the new Czech general management plan is based on it).

Cooperation on habitat management
General management plans of both the NPs were widely discussed and consulted, especially in NP board meetings in which representatives from the other park take part. More than 80 per cent of the total area of the BR is covered in forests, and all of them were affected by air pollution in different degrees of damage, especially in the 1970–90s. Most of the monitoring, research and management activities and intensive international cooperation were therefore focused on the protection of forest stands. Forest management, including hunting issues, has been the responsibility of the Polish NP since the foundation of the park, and of the Czech NP since 1994. In recent years,

Top: Bilateral staff training (© Kamila Antosova).
Bottom: Meeting of park directors at the state border between Poland and Czech Republic (© Jiri Dvorak).

the exchange of experience among Czech and Polish forest managers as become the rule. Such exchanges take place e.g. in the course of field excursions organised in forests of the neighbouring NP, workshops devoted to problems of forest management in NPs, sharing scientific knowledge at bilateral conferences etc., and they are very useful, even if practical approaches and opinions sometimes differ (in particular with regard to different approaches to management of polluted forests or bark-beetle outbreaks in NP core zones). Wide-ranging consultations have taken place with Polish colleagues on the newly prepared Czech forest management plan for the next decade (based on the chapter relating to the general management plan, written in line with the Polish plan for the conservation of forest habitats, coordinated forest habitat evaluation on both sides of the border, comparable forest typology used etc.). Consultation and coordination has also taken place with regard to other management approaches (especially in respect of tourism and recreation) between the two NPs recently – e.g. on connectivity of bike trails crossing the state border, joint approaches to illegal alpine skiing activities and winter mountaineering in core zones etc.

Cooperation on landscape and species protection
Once a common need for landscape or species protection is determined, both national park authorities try to develop

a national project and submit it for the purpose of financing (either separately or jointly – based on national legislation or conditions of funds applied for). If the proposal is accepted on both sides, it is run in parallel with regular contacts between the two NPs and, as far as possible, according to the same guidelines. An example for nature conservation is the data collection for habitats and species of European concern for the EU Natura 2000 Network. Whereas the applied methodology was coherent on both sides of the border, the results were submitted separately to the EU according to the relevant national requirements. Habitat conservation is the best approach to maintaining biological diversity, but in some cases it is necessary to focus more precisely on individual species. In the Czech NP, the most endangered plant species with limited populations are protected 'ex situ' in a botanical garden ('gene-bank'). Seeds of some of the species are collected from wild plants, cultivated 'ex-situ'. The seedlings raised are reintroduced in their original habitat. This is another example of cooperation between the two NPs, e.g. the only population of Saxifraga nivalis is known to exist in the Polish glacial corrie and Czech botanists reintroduced seedlings – cultivated in the Czech 'gene-bank' – in the Polish corrie.

Cooperation on research and monitoring

Without the support of logistics, it would be impossible to carry out any biodiversity conservation at all. A new 'Concept of monitoring and research in the Krkonose NP' has been prepared for the next decade. Again, it is based on the general management plan, and it was discussed, commented and agreed in Board meetings of the Czech NP, in which the Polish NP is represented, and the main issues inherent in the concept (e.g. coordination, databases, methodologies, presentations, various projects etc.) are taken bilaterally to the Czech-Polish level. Basic inventory surveys of flora and fauna are completed and/or updated on both sides of the mountains and data is exchanged regularly. Both Czech and Polish scientists cooperate on joint projects which cover the area of the entire BR (e.g. mapping of breeding birds distribution, with results published in a bilingual book) for the most valuable localities on both sides along the border (e.g. the international GLORIA project). The preparatory work consisted in establishing an inventory of vital importance for inclusion of Krkonose/Karkonosze in the Natura 2000 Network. Joint Czech-Polish projects of both NPs on visitor monitoring in NP core zones and on telemetry of red deer along the state border are presently in preparation, with applications by both NPs for EU transboundary finance pending. The regular Czech-Polish scientific conference on 'Geoecological Problems of the Krkonose/Karkonose' has by now become an established tradition as an occasion where scientists who work in the entire mountain range exchange observations. Starting in 1991, a total of seven conferences have been organised so far at three-year intervals, hosted alternately on the Czech or Polish side of the mountains. Likewise, transboundary workshops under the title of 'Cloudberry' are organised annually for university students involved in bachelor or diploma theses in the Czech or Polish Krkonose Mountains.

Important steps have been taken in terms of data management and application of geographic information systems for planning and data evaluation (e.g. preparation of joint GIS layers for the entire mountain range). Both NPs are really keen on the exploitation of GIS, and several joint projects to improve its utilisation have either been completed or are under way (such as the joint project with UNEP GRID Warsaw). The aim is harmonise the scale of geographical data available for both NPs and to harmonise interpretation (standardising the interpretation methodology applied on the Czech and Polish sides). The level of Czech-Polish cooperation and coordination in logistics activities has improved rapidly in recent years, and joint projects run by Czech and Polish scientists are beginning to become the norm. It is encouraging to note the decline in the amount of lectures and papers (including figures and references) that maintain the (formerly cherished) concept that there are no northern (for Czechs) or southern (for Poles) slopes to the mountains.

Participants of a bilateral conference on 'Geoecological problems of Krkonose/Karkonosze' (© KRNAP).

Cooperation on education and public relations

Educational activities include a wide spectrum of different information/cultural facilities such as
* Krkonose Environmental Education Centre,
* an ecological exhibition on 'Rocks & Life' on the Czech side,
* the Karkonosze Centre for Environmental Education in Poland,
* programmes and courses open to children and adults, visitors and local people; several of them are especially for younger generations (e.g. a new series of programmes 'Learned from Nature' was started by the Czech NP in 2010),
* the monthly popular journal 'Krkonose – Jizerske hory' published by the Czech NP but with Polish contributions and a Polish NP representative on its editorial board,
* Opera Corcontica, a yearbook of scientific papers prepared by the Czech NP but in close cooperation with staff from the Polish NP and scientists from the Polish side, etc.
* attempts to raise the level of ecological thinking in the population, including the awareness of importance of biological diversity. Transboundary cooperation in Krkonose/Karkonosze is very productive in this field.

In recent years, we managed to run a number of joint projects in the field of environmental education. One of the first joint activities was the installation of information points operated by touch screens with information on the mountains, the local wildlife and biodiversity heritage and its protection. Karkonosze Environmental Education Centre in Szklarska Poreba (PL) is one of many outcomes of those projects, and so are numerous specific educational tools and materials. Another joint Czech-Polish project, with participation from the Krkonose NP museum and financed by EU funds, is now under way. It is called 'Via Fabrilis' with a general focus on local crafts in the Krkonose region, on its local history of crafts and on the support of the existing local craftsmanship. Dozens of staff members from both NPs are now starting to learn the native language of their colleagues on the other side of the border. The two NPs also cooperate on the Junior Ranger Project and on school exchanges.

Concluding remarks

The efforts made by the two national park authorities in stimulating cooperation in the Giant Mountains had their reward in 2004, when Krkonose/Karkonosze was certified by EUROPARC Federation as an exemplary Transboundary Park. The certificate is currently being re-evaluated by the Federation. It is safe to assume that the establishment of the Transboundary Biosphere Reserve in 1992 substantially contributed to this success. Although it does not constitute one coherent body, it provides an open forum for communication between all stakeholders. The concept of the TBR as a forum has the advantage that it is free from the structural difficulties (e.g. differences in legislation, financial resources, administration and hierarchy) which affect the two sides. The good level of cooperation between the two parks has allowed them to take a stronger stand against inappropriate development projects such as proposals for ski lift construction or new roads.

References

Fall J. & Jardin M. (eds) (2003). Five transboundary biosphere reserves in Europe. Biosphere Res. Tech. Notes, UNESCO Paris: 1–95.

Flousek J. (1996). Cooperation in biodiversity conservation in the Czech and Polish Krkonoše national parks and biosphere reserve (Krkonoše and Karkonosze). In: Hamilton L. S., Mackay J. C., Worboys G. L., Jones R. A. & Manson G. B. (eds): Transborder protected area cooperation. Australian Alps LC, IUCN: 31–38.

The partnership agreement being signed in 2007 by (from left) Mr Hawrysh (Redberry Lake BR), Mr Abe (Rhön BR) and Ms Pollock (Georgian Bay Littoral BR). (© R. Braun).

International Partnerships and Learning Platforms: The Cooperation between a German and Several Canadian BRs

by Rebecca Pollock, Karl-Friedrich Abe, Reinhard Braun, Andrew Hawrysh & Claude LeTarte

As models of community-based conservation and sustainable development, UNESCO biosphere reserves are intended to share their experiences with other communities internationally. Together, they form the World Network of Biosphere Reserves, linked by a common understanding of purpose. Networking is achieved by exchanges of ideas, experience and people at all levels. In this way, biosphere reserves (BRs) create 'learning platforms for sustainable development'.

In 2005, a partnership was established between the Thuringian part of Rhön BR in Germany and three biosphere reserves in Canada. A signed partnership agreement formalises long-term cooperation, knowledge transfer and a joint implementation of UNESCO's objectives for the MAB Programme. Although each of the BRs contain highly diverse landscapes and conduct a wide range of regional activities themselves, their models of sustainable development are comparable as they strive to establish 'Quality Economies' in the areas of sustainable tourism, agriculture, and conservation (supported by research and monitoring). The following case study illustrates an experience and knowledge-transfer between the management and staff of the Rhön, Charlevoix, Georgian Bay, and Redberry Lake BRs as well as the development of specific joint products, such as tourism publications and training workshops.

Development of the cooperation

The regional network of biosphere reserves for Europe, North America and Israel is known as the EuroMAB network. Meetings are held every two years to exchange experiences among BR managers, national MAB committee members, as well as researchers and scientists. In 2005, the meeting was hosted by Austria in the Wienerwald BR which offered the possibility for members to get to know each other and to initiate partnerships within the worldwide UNESCO Network. In 2006, Mr Abe was invited to the annual meeting of the Canadian Biosphere Reserves Association (CBRA) in Redberry Lake BR (Saskatchewan). Together with Mr Reinhard Braun, a GIS expert, the Rhön BR was introduced in a presentation showing the management aims of this BR and specific examples, such as hiking trails promotion (a premium hiking trail called 'DER HOCHRHÖNER') and regional product development. Following the meeting, the German party toured four other BRs. At the Niagara Escarpment BR, they were interested in planning processes and visitor services. In the Georgian Bay BR, the focus was on sustainable and ecotourism development and communication of natural and cultural heritage values. They crossed the province of Ontario by bus (through Algonquin Park) to the national capital of Ottawa. The next day, the German party met with members of the Frontenac Arch BR situated on the St Lawrence River to learn about local agricultural product marketing. Finally, they took a train to the city of Montreal and the nearby Mont St-Hiliare BR to learn about public participation in that biosphere reserve. Their visit included five Canadian BRs in only ten days!

In October of the same year, the German hosts welcomed Canadian BR representatives Andrew Hawrysh (Redberry Lake BR), Rebecca Pollock (Georgian Bay BR) and Charles Roberge (Charlevoix BR) to spend five days touring the Rhön region. The goal for this exchange was to build capacity and share ideas, projects and strategies for sustainable development. The following includes a list of workshop themes presented to the Canadian participants:
- Eco-tourism development,
- Product labelling and quality economies,
- Landscape conservation and land use conflicts,
- Agriculture, organic farming and marketing,
- National and international partnership development.

They learned about the role of small businesses in using and marketing local products, such as wood and wool. They also heard presentations on 'biosphere reserve labelling and branding' and how a BR logo is used on products ranging from organic milk and meat to hotels and restaurants. The Canadian party also visited a brewery and a wood factory, all of which have a role in supporting traditional plants from the region, while providing local employment. They made presentations to local and provincial officials, including the Minister of Environment. They met with students and teachers at a local high school and outdoor education centre, and had television and newspaper interviews with the media. Their discussions with their German hosts gave them information on new BR management approaches.

Lessons Learned

Canadian biosphere reserve representatives were impressed by the political support given to German biosphere reserves, the innovations used to stimulate regional quality economies, and the educational activities, public awareness and local pride in being part of a UNESCO world biosphere reserve. Specifically, they noted:

- Political Leadership: how biosphere reserves are supported at all levels of government – federal, state and local level.
- Innovation: economic development that occurs in the Rhön is guided by BR principles, especially the use of regional products, quality labelling and marketing; new entrepreneurs are encouraged to create small businesses using local products.
- Integration: all of the biosphere reserve activities seem to be working in harmony. Villages and towns have signs with a common logo; hotels and restaurants display the quality label; children are taught environmental education and students work on ecology projects that directly involve the community.

German biosphere reserve representatives from the Rhön were interested in the strong level of public participation and the 'multi-stakeholder' management models of Canadian BR organisations. In Canada, each biosphere reserve has a unique governance structure. Some are affiliated with research institutions, others work closely with national parks, and others are set up as independent, non-governmental organisations that rely on project funds and private donations. In particular, it was evident that public participation was critical to BR success:

- Charters for Sustainable Development are a statement of values that are developed with local businesses, cultural organisations and the tourism industry. They help to define sustainable development for a particular region and engage individuals and organisations to adopt the charter to improve their environmental performance. In Canada, four biosphere reserves have created charters: Charlevoix, Lac St Pierre, Fundy and Frontenac Arch.
- The Decade on Education for Sustainable Development is supported by individual BR and their education programmes. Several BRs participate in the UNESCO Associated Schools Programme. In others there is school curriculum developed specific to the biosphere reserve concept and networks set up to support educators and partners (e.g. museums, parks, natural and cultural history groups).
- The Canadian BR Association is a non-profit, non-governmental organisation that supports Canadian BRs in the achievement of their UNESCO mandates and demonstrates their collective value nationally and internationally. Through CBRA, biosphere reserve managers, project coordinators,

and volunteers can maintain communications among themselves and with other related organizations, collaborate on shared projects, and exchange local expertise among biosphere reserves in Canada and with biosphere reserves around the world.

Following this first experience of professional exchanges and learning tours in both countries, the partners developed their interest in continued cooperation and signed a partnership agreement at the 2007 annual meeting of CBRA held in Georgian Bay. Since the 2007 EuroMAB meeting in Antalya (Turkey), the partnership has included the development of a tourism brochure in three languages promoting the biosphere reserves in both countries and specific attractions in the four sites. In 2008, German representatives presented a workshop on quality economies and product labelling for CBRA members and attended the annual meeting of CBRA in Mount Arrowsmith BR. In 2009, Canadian delegates travelled through the Thuringian region of the Rhön during the 'Year of the Biosphere Reserve' in Germany. Discussions were held with local administrations about proposals for youth exchanges, the need for an international workshop on sustainable agriculture, biodiversity and local food, as well as rural community adaptations to climate change. In 2009, the Tatry biosphere reserve in Stara Lesna (Slovakia) hosted 115 delegates from 22 different countries at the EuroMAB meeting. The group worked on a Strategic Action Plan to improve online communication with a formal web-based learning platform, share governance models, set up partnership projects – like school twinnings – and learning exchanges. Delegates shared activities from their regions and made recommendations for conservation research, for sustainable development projects and for environmental education.

Conclusion

The international partnership that has developed between one German biosphere reserve and several in Canada is an example of the value of the MAB Programme coordinating regional networking and knowledge exchange, particularly as individual biosphere reserves receive different levels of support nationally and locally. Exchanges between biosphere reserve managers allow them to see the MAB Programme operating in a different culture and context, compare approaches and transfer effective programme ideas.

Sustainable development will be defined differently by each community involved but common values and participatory approaches are key to their success. Biosphere reserves have a facilitation role in their regions to involve diverse stakeholders and create social networks that support biodiversity conservation, local livelihoods, and regional economies. They accomplish this through education, leadership and innovation, striving to become UNESCO's ideal of 'learning platforms for sustainable development'.

Rhön Biosphere Reserve (1991)

This biosphere reserve is situated in the low mountain ranges of the Rhön in the centre of Germany. In contrast to other German low mountain areas, the Rhön is also known as the 'land of open vistas' representing an open cultural landscape shaped by human use for many centuries. Naturally, the region would be covered by beech forest (*Fagus sylvatica*), however extensive farming and dairy cow raising transformed forests mainly into montane and sub-montane humid grasslands on siliceous soils. Two bogs host numerous endangered animal and plant species.

The Rhön was designated as a biosphere reserve after the reunification of Germany covering three Länder (federal states) – Thuringia, Bavaria, and Hessen. Each of the three regions has their own management offices. A framework management plan for the protection, maintenance and development of the Rhön Biosphere Reserve has been elaborated with the participation of all stakeholders and includes the conservation of agricultural biodiversity, the branding and labelling of quality products, and the promotion of nature tourism and hiking trails.

About 162,000 inhabitants live in this rural area. Apart from agricultural activities, people make their living from small businesses and tourism. Partnerships among hotels, restaurants, farmers, and artists seek to link all activities in the BR. The Rhön is known for direct marketing of regional products. For instance, products from the Rhön sheep, an endangered breed adapted to the rough Rhön climate, and apple products from regional orchards are marketed. Several visitor centres have been established providing diverse environmental education programmes to the public.

Top: A local shepherd in Rhön BR, proud of his herd, his dogs and the landscape; bottom: Promotion of local products (© Abe).

Impressions of Charlevoix BR, Canada (© Claude Letarte).

Charlevoix Biosphere Reserve (1988)

Situated 80 km east of Quebec City, Charlevoix BR borders the Saint Lawrence River to the north. A majestic landscape of mountains and sea, forests and shores, the Charlevoix region includes the drainage basins of the Malbaie River and the Rivière du Gouffre. It covers a total area of 560,000 hectares stretching from Petite-Rivière-Saint-François to Port-au-Saumon in the east and from the middle of the St Lawrence River (including Île-aux-Coudres) to the Réserve faunique des Laurentides in the north. Extending from 5 to 1,150 metres above sea level, the area comprises agricultural areas, river ecosystems, estuarine tidal marshes and flats, coniferous and mixed forests, stunted vegetation (Krummholz) and mountain tundra ecosystems.

Approximately 30,000 people live in the area, in over a dozen municipalities and small towns. In former times, the population of Charlevoix used to rely on the river and the sea, for example on coastal navigation, marine constructions and fisheries (e.g. beluga whales). Today, the economic landscape has diversified and major factors in the local economy are now forestry, silica mining, agriculture and tourism. The forest education centre 'Les Palissades' or the ecological centre 'Port-au-Saumon' are important institutions for environmental education in the area. The BR team participates in the development of regional prosperity and community pride and now cooperates in the creation of the 'Institute Hubert-Reeves for Science and Research'. Current activities include the campaign 'Towards a Sustainable Landscape', which promotes eco-tourism, cultural heritage, and green business practices supported by the development of a Charter. Environmental education themes include: energy and water conservation, sustainable forestry and agriculture, and biodiversity protection. Cooperative conservation activities focus on monitoring and restoring the health of hydrological basins and forested areas.

Redberry Lake Biosphere Reserve (2000)

Redberry Lake BR is situated in the province of Saskatchewan in the south-west of Canada, covering 112,200 hectares. The regional landscape is composed of rolling prairie, dotted with seasonal ponds and marshes, along with aspen/shrub groves. The core area is a saline lake with several islands. There are small patches of natural mixed prairie which is very rare in this highly grazed and cultivated part of the prairies. Redberry Lake is an important site for the conservation of several significant species of birds. It provides habitat for nine endangered, threatened, or rare bird species, as well as over 180 other species. Monitoring nesting sites of American White Pelican (*Pelecanus eryhthrorhynchos*) is one of the research and monitoring activities undertaken in the area.

Only about 1,000 people live in this rural area and most of them are Euro-Canadians, primarily of Ukrainian origin. The primary economic activities in the region are agriculture and livestock breeding. Eco-tourism development over the past decades has encouraged new ways of looking at local habitats, and brought new hope to some community enterprises, including sustainable tourism (wildlife viewing) and organic agriculture. There exists a strong potential to undertake the development of more sustainable agriculture, livestock, and silviculture products that could be marketed under the 'brand' of the biosphere reserve, such as 'model' farms and natural prairie grass cultivation for seed stocks. There is also a potential for linkages with other biosphere reserves to market the products of sustainable resource use, as well as educating the general public about conservation practices.

Typical landscape of Redberry Lake BR, Canada (© Andrew Hawrysh).

Top: Canoeing in Georgian Bay Littoral BR; bottom: Winter shoreline (both photos: © Kenton Otterbein)

Georgian Bay Littoral Biosphere Reserve (2004)

This biosphere reserve encompasses the largest island archipelago of the North American Great Lakes. Known locally as 'the 30,000 Islands', it is a complex association of bays, inlets, sounds, islands and shoals lying along the edge of the Canadian Shield bedrock which rises as low lying hills and ridges on the adjacent mainland. This topography supports a rich mosaic of forest, wetlands, and rocky habitat types with associated biodiversity.

It also has high scenic values which attract large numbers of summer residents, cruising boaters, and seasonal visitors. The number of permanent residents associated with the biosphere reserve is about 17,000, but summer residents and visitors increase this some 3 to 5 times more, and up to 25 times in some more accessible localities. Most of the area is accessible only by boats. The main development issues are promotion of best practices, especially for water-oriented recreation and ecotourism linked to particular destinations.

Within the 347,000 hectares, the core area is made up of one national park and five natural environment or nature reserve provincial parks. The buffer zone is composed of 14 provincial Conservation Reserves, and the core and buffers together form a contiguous landscape unit along the eastern Georgian Bay coast. The inaccessibility of much of the transition area adds to the conservation function. Administration and management are provided by the non-governmental organisation, Georgian Bay BR, Inc. which represents a number of stakeholder interests which together coordinate the BR programme.

Retezat National Park and Biosphere Reserve in Romania (© Cristian-Remus Papp).

Tracking Management Effectiveness: Experiences from two Carpathian Biosphere Reserves

by Cristian-Remus Papp

The evaluation of management effectiveness is defined by IUCN's World Commission on Protected Areas (2006) as the assessment of how well protected areas (PAs) are being managed – primarily the extent to which they are protecting values and achieving goals and objectives. The term 'management effectiveness' reflects three main 'themes' in protected area management:

- design issues relating to both individual sites and protected area systems;
- adequacy and appropriateness of management systems and processes; and
- delivery of protected area objectives including conservation of values.

The number of PAs has increased exponentially for over a century (now there are more than 138,000 PAs registered in the World Database on Protected Areas) and the tendency remains the same, owing to conservation efforts in most parts of the world. Paradoxically, the 'Living Planet Index' (2010) shows a decline of about 30 per cent for 2,544 mammal, bird, reptile, amphibian and fish species between 1970 and 2007. This means that the efforts and all kinds of resources invested in establishing and managing PAs are generally not yet seen to reap rewards.

The responsibility for the effectiveness of PAs and their contribution to biodiversity conservation and poverty alleviation rests mainly with the management (of both individual PAs and PA systems). However often PA managers, with their multiple tasks and activities on site, hardly have time to think and reflect on the importance of assessing their management effectiveness. However, the Programme of Work on Protected Areas (PoWPA), adopted by the 7th CBD Conference of Parties in 2004, stressed the importance of 'evaluating and improving the effectiveness of protected areas management' (Dudley et al, 2005). The management task becomes more complex in the case of biosphere reserves (BRs), where biodiversity conservation and sustainable development have to be addressed considering the local communities' needs and desires, and applying sound science at the same time. The multitude of tasks to be performed by BR managers requires very good planning, and knowledge and skills in the fields of organisation, and especially adaptive management. Adaptive management in the case of BRs should be an approach to environmental management, also taking into account the context of complex economic and social systems, to optimise decision-making in the face of uncertainty, by identifying the various uncertainties over time using system monitoring. To optimise and improve the efficiency of BRs, managers have first to assess their past activities and achievements, analyse and understand the present context and situation in order to maximise the use of available resources in the future and establish systems for monitoring the status and trends of PAs and their values. The best and probably easiest way to start is the use of a comprehensive assessment system or tool. The following paragraphs try to show how the Carpathian Protected Areas Management Effectiveness Tracking Tool may help PA Managers to assess and improve management practices.

The CPAMETT

Due to the differences in PAs (e.g. terrestrial or marine, national or international category, designated for species or landscape conservation), several methodologies and tools have been created to assess PA management effectiveness and to verify the achievement of the objectives for which PAs were established. Most of the methodologies and tools for assessing PA management effectiveness are either based on the IUCN-WCPA Management Effectiveness Evaluation Framework, or take this guideline into account. The Framework is a guide for developing comprehensive assessment systems and based on six elements: context, planning, inputs, processes, outputs and outcomes. These elements are important in developing an understanding of how effectively PAs are being managed. They reflect three large 'themes' of management: design (context and planning), appropriateness/adequacy (inputs and processes) and delivery (outputs and outcomes) (Fig. 1).

The European study on the management effectiveness evaluation of PAs, performed by the Universities of Greifswald

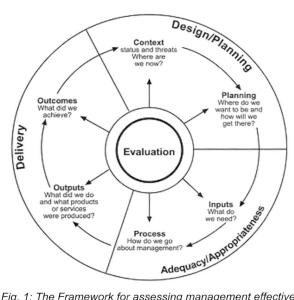

Fig. 1: The Framework for assessing management effectiveness of protected areas (Hockings et al. 2006).

and Queensland, in partnership with the UNEP World Conservation Monitoring Centre, EUROPARC Federation and the German Federal Agency for Nature Conservation, revealed that there are about 40 different approaches to PA management effectiveness evaluation in European countries (Nolte et al. 2010). One of these is the Carpathian Protected Area Management Effectiveness Tracking Tool (CPAMETT). It was developed within the '2012 Protected Areas for a Living Planet Programme' (2012 PA4LP), which is a global programme initiated by WWF to promote and support the implementation of the CBD PoWPA. The PA4LP is implemented in five priority ecoregions: Carpathians, Dinaric Alps, Caucasus, Altai-Sayan and West Africa Marine. The CPAMETT is an advanced version of the Management Effectiveness Tracking Tool (METT) which was developed by WWF and the World Bank. It aims at monitoring the progress in the performance of PAs in the Carpathian countries. It is an online, web-based tool and consists of two major components:

- Two forms for collecting the information on protected areas (Info on my Protected Area) for assessing the management effectiveness (Assessment Form) and the results section;
- A database on protected areas of the Carpathian region.

Using the tool is relatively easy. First of all, PA managers or administrators have to register and enter the basic information on their site (e.g. name, category, size etc). Subsequently, the 'Assessment Form' has to be completed. PA Managers have to answer 42 questions relating to their management performance. Some of the original METT questions (30) have been split and/ or slightly adapted, for instance cultural and natural values were separated. Accordingly, some answers were changed to a more quantitative manner in order to enhance objectivity. For each question, PA managers have to choose only one answer out of the four possible, which is scored from 0 to 3, depending on the answer. For some of the questions or elements, additional points can be obtained. For instance, in the case of 'management plan', up to three additional points can be obtained if:

- the planning process allows adequate opportunity for key stakeholders to participate in the establishment and periodic review of the management plan, and to participate in influencing the management plan, as long as this is not to the detriment of the protected-area objectives;
- there is an established schedule and process for periodic reviews and updating of the management plan; or if
- the results of monitoring, research and evaluation are routinely incorporated into the planning.

A maximum of 156 points can be achieved overall. Comments and proposed actions to address specific problems can be added to each question. After the assessment, the degree of effectiveness can be checked under the 'Results' section. There are different possibilities to view the results, e.g. sorted by the relevant IUCN-WCPA element (see details above), for each individual question, compared with average scores of individual Carpathian countries or the Carpathian region etc. Using the WCPA elements, a graph can be generated to compare the present situation of the PA to the potential ideal. Results of repeated assessments can also be compared with the results of previous years, and improvement or decline of performance can be detected easily. Several filters can be used for the purpose of comparing results; for instance, one can select a specific PA category (national, IUCN, international) or the size of the PA. Moreover, comparisons can be made at different levels (within the Carpathians of a specific country, within a specific country, within the Carpathian Mountains Ecoregion or within all seven Carpathian countries). The various filters available can help PA Managers to narrow down their search (Fig. 2).

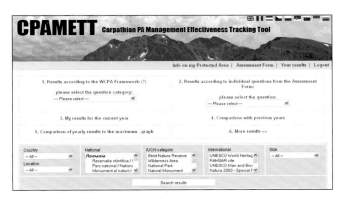

Fig. 2: CPAMETT results section.

It is important to mention that both CPAMETT (with the 42 questions) and METT (with the original 30 questions) reports can be generated. In the second part of the tool (on-line database), reports can be generated on relevant topics from the first form and on any individual question from the assessment form. Here as well, comparisons can be made at different levels (as in the results section of the first part). The reporting features can help PA managers to establish contacts and share information or experience on project implementation, based on the information provided in the first form. The CPAMETT was translated into all seven Carpathian languages, and also into Bulgarian (there are also other countries interested in using the tool). The availability of the tool in the national languages is essential for being used by the greatest possible number of PAs. The tool was tested in the period between May 2009 and June 2010. More than 50 PAs have been assessed so far in five different Carpathian countries (Romania, Hungary, Czech Republic, Slovakia and Serbia). Workshops, as well as demonstration and training sessions have been organised with representatives from the Ministries of the Environment, nature conservation agencies, PA administrations and NGOs in six countries (the five listed above plus Poland) on the use of the tool and its importance in improving management effectiveness. Additionally, a direct link to CBD PoWPA will be created to allow governments to report back to the CBD in an easy and effective way.

Strengths and weaknesses of CPAMETT

Strengths:

- Comprehensive on-line tool, easy to use;
- Very useful tool for reporting to the CBD and for fulfilling the obligation to achieve, by 2010, the assessment of at least 30 per cent of the PAs of a signatory party;
- Easy to analyse the results and to generate different types of reports;
- The collected data is stored in a database thus reducing the amount of paperwork to be done;
- Provides an opportunity to compare the results of PAs within one country (at national level) or a region (within the Carpathian region of a specific country);
- Facilitates making contact, finding different experts and sharing information or experience on project implementation;
- Internationally embedded links to the CBD, UNEP World Conservation Monitoring Centre (WCMC) and the World Database on Protected Areas (WDPA) – the tool was taken up by UNEP-WCMC with some modifications and is now used in Asia for the Tiger Conservation Landscape PAs.

Weaknesses:

- It is possible that only one person performs the evaluation (e.g. no internal discussion takes place). It is recommended that the forms are filled in by the entire PA staff, having internal discussions and debates. In addition, other experts and stakeholders can be involved as well.
- Depending on the PA staff, the evaluation can be subjective.
- Lack of institutionalisation.
- If the internet connection is not reliable, it is recommended to use printed forms for back-up purposes (in case the forms are not saved on-line from the beginning, the data can be transferred to the database subsequently).

The assessment of biosphere reserves in the Carpathians

So far, the CPAMETT was tested in nine biosphere reserves in the Carpathian Ecoregion, namely in the:

- Retezat and Pietrosul Rodnei (as part of the Rodna Mountains NP) BRs, both in Romania;
- Palava and Bile Karpaty BRs in the Czech Republic;
- Tatry, Slovenský Kras, Polana & Vychodne Karpaty in Slovakia;
- Djerdap NP (as part of the Iron Gates – Djerdap BR).

The assessments of the Retezat and Pietrosul Rodnei BRs were performed with the involvement of all members of the management staff on both sides. Actually, this is the idea behind the use of the tool – to bring together the management team for some productive discussions that can form the basis for effective adaptive management. This can engage the team in identifying issues and even in 'brainstorming' to find solutions or to determine any necessary steps to resolve specific problems. In both BRs, it turned out that the CPAMETT can raise questions which some managers have not even thought about. In the first form, especially the appreciation and ranking of threats was found very useful by managers and teams. They had to discuss the twelve major categories of threats (as classified by IUCN) – from residential and commercial development within the PA to specific cultural and social threats – and tick them as either of high, medium, low significance or not applicable. Some discussions took place regarding the perception of the level of threats, for instance in the case of hunting, logging and fishing within the biosphere reserves, but also with regard to recreational activities. Some of the threats attracted special attention and triggered debates. It was interesting to see that in some cases, the final answer was given by rangers, who were better informed about the situation in the field. For instance, deliberate vandalism and destructive activities were perceived differently by individual staff members. The rangers who knew these aspects very well were to some extent contradicted by one of the managers who

Fig. 3: Stakeholder involvement and discussions amongst staff members are quite important for evaluating management effectiveness (© Cristian-Remus Papp).

Fig. 4: Scenic landscape in Iron Gates – Djerdap BR, Serbia – Romania (© Cristian-Remus Papp).

– owing to the overwhelming administrative issues – could not manage to see these negative activities in the field or was not sufficiently updated on what went on. On this occasion, the team members learned from each other and became much better acquainted with different issues. When completing the two forms, the teams realised the importance of having all members present. The staff with responsibility for biodiversity, community outreach and field work are in possession of specific knowledge and key information which gives them a deep understanding and enables them to come up with the most appropriate answers. For example, when discussing pathogens, invasive species, conservation or the monitoring of key indicator species and habitats, the most reliable person in providing this type of information is the biologist of the PA. The discussions on this wide range of management tasks and issues demonstrated the complexity of the subject of managing biosphere reserves and where the weaknesses lie. The assessments also highlighted that more human, material and financial resources would be required in order to achieve more effective management. Capacity building and stakeholder involvement and participation are needed especially in the case of the Pietrosul Rodnei BR. Thus, in the course of completing the assessment form, some of the most stringent aspects of a PA and its management were dealt with in detail and analysed. Activities started to be planned and were written down in order to resolve some of the problems. The great advantage of this process was that all members were involved in the analysis of different problems and were able to contribute their own opinions and ideas. Moreover, the meeting allowed them to recognise jointly what changes should be planned for the future. After performing the assessment, the team was able to visualise the results and pinpoint precisely where the strengths and weaknesses are. The more tangible method of viewing the results provided a more complete overview of management effectiveness. The BR team was able to identify the situation – an important step towards finding ways to improve it. It was a very practical and useful exercise for staff members, which demonstrated that a participatory approach in the assessment of management effectiveness can generate solutions and answers to complex management issues which are then available for immediate implementation.

Comment of Zoran Acimov (Director of Retezat National Park administration, Romania), who used the CPAMETT to jointly evaluate the management effectiveness of Retezat National Park with his team

The use of such a tool was and still is very necessary in order to be able to evaluate in a reasonable way the status of a protected area or of a network of protected areas. The questionnaire of the CPAMETT is very comprehensive and precise (it might be too precise in some aspects; i.e. geographical coordinates are not too relevant for large PAs). It gives a clear image of the current status/condition of the protected area, and can be used as a starting point in taking some decisions in the management process. Meantime, it very much depends of the accuracy of answers and the honesty of those who give the answers. There are only few external verification possibilities of the data which is introduced in the tables. Doing it in a consultative manner with the whole management team might increase the objectiveness, but still in small groups the leader can somehow "guide" the group's opinion.

Regarding myself, I have accepted from the beginning the use of that tool in the process of management effectiveness of our park, even if Retezat is not an typical biosphere reserve but rather a very clear example of a national park, corresponding to Category II of IUCN.

References:

Dudley, N., Mulongoy, K.J., Cohen, S., Stolton, S., Barber, C.V. & Gidda, S.B. (2005). Towards Effective Protected Area Systems. An Action Guide to Implement the Convention on Biological Diversity Programme of Work on Protected Areas. Secretariat of the Convention on Biological Diversity, Montreal, Technical Series no. 18.

Hockings, M., Stolton, S., Leverington, F., Dudley, N. & Courrau, J. (2006). Evaluating Effectiveness: A framework for assessing management effectiveness of protected areas. 2nd edition. Best Practice Protected Area Guidelines Series No. 14. IUCN, Gland / Cambridge.

Nolte, C., Leverington, F., Kettner, A., Marr, M., Nielsen, G., Bomhard, B., Stolton, S., Stoll-Kleemann, S. & Hockings, M. (2010). Protected Area Management Effectiveness Assessments in Europe: A Review of Application, Methods, and Results. Publisher BfN, Federal Agency for Nature Conservation, Bonn, Germany.

WWF, Global Footprint Network & Zoological Society of London (2010). Living Planet Report 2010. Biodiversity, biocapacity and development.

Editor

Austrian MAB Committee, Austrian Academy of Sciences,
Dr.-Ignaz-Seipel-Platz 2, A-1010 Vienna

Editorial staff

Technical concept: Sigrun Lange (E.C.O. Germany)
Project coordination & Technical editing: Sigrun Lange
English proof-read: Brigitte Geddes (Freelance translator)

Layout

Sigrun Lange (http://www.e-c-o-deutschland.de)

Participating authors

Abe, Karl-Friedrich (Rhön BR, Germany)
Artemov, Igor (Katunskiy BR, Russian Federation)
Aspizua, Rut Cantón (Sierra Nevada BR, Spain)
Badenkov, Yuri (Russian Academy of Science, Russian Federation)
Bender-Kaphengst, Svane (NABU, Germany)
Bertzky , Monika (UNEP-WCMC, UK)
Bomhard, Bastian (UNEP-WCMC, UK)
Bonet García, Francisco (University of Granada, Spain)
Borowski, Diana (Centre for Mountain, Scotland)
Boussaid, Mohamed (GTZ, Morocco)
Braun, Reinhard (Rhön BR, Germany)
Cano-Manuel, Javier (Sierra Nevada BR, Spain)
Dovhanych, Yaroslav (Carpathian BR, Ukraine)
Effler, Dirk (Effler Consulting, Germany)
Flousek, Jiri (Krkonose NP, Czech Republic)
Grabherr, Georg (University of Vienna, Austria)
Gubko, Victoria (Carpathian BR, Ukraine)
Hamor, Fedir (Carpathian BR, Ukraine)
Hawrysh, Andrew (Redberry Lake BR, Canada)
Henares, Ignacio (Sierra Nevada BR, Spain)
Kalmikov, Igor (Altaiskiy BR, Russian Federation)
Kašpar, Jakub (Krkonoše NP, Czech Republic)
Köck, Günter (Austrian Academy of Sciences, Austria)
Lange, Sigrun (E.C.O. Germany)
LeTarte, Claude (Charlevoix, Canada)
Maikhuri, R.K. (G. B. Pant Institute of Himalayan Environment and Development, India)
Messerli, Bruno (Switzerland)
Munteanu, Catalina (Centre for Mountain Studies, Scotland)
Nune, Sisay Hailemariam (NABU, Ethiopia)
Papp, Christian-Remus (Freelancer, Romania)
Pedraza Ruiz, Roberto (Sierra Gorda BR, Mexico)
Pollock, Rebecca (Georgian Bay BR, Canada)
Pokynchereda, Vasyl (Carpathian BR, Ukraine)
Price, Martin (Centre for Mountain Studies, Scotland)
Rao, K.S. (Delhi University, India)

Rodríguez-Rodríguez, David (Spanish National Research Council)
Sánchez, Javier (Sierra Nevada BR, Spain)
Sarmiento, Fausto (University of Georgia, USA)
Saxena, K.G. (Jawaharlal Nehru University, India)
Schaaf, Thomas (UNESCO, France)
Schmidt, Matthias (Freie Universität Berlin, Germany)
Shigreva, Svetlana (Altaiskiy BR, Russian Federation)
Thiel, Lydia (Teacher, Germany)
Vološčuk, Ivan (University of Matej Bel, Slovak Republic)
Yashina, Tatjana (Katunskiy BR, Russian Federation)
Zamora, R. (University of Granada, Spain)
Zisenis, Marcus (ECNC, The Netherlands)

Photo credits

Title, top: © Sigrun Lange
Title, bottom, left: © Lammerhuber / Planet Austria
Title, bottom, right: © Svane Bender-Kaphengst
Chapter 1: © Sigrun Lange
Chapter 2: © Sigrun Lange
Chapter 3-1: © Lammerhuber / Planet Austria
Chapter 3-2: © Sigrun Lange
Chapter 3-3: © Sigrun Lange
Back: © Harald Pauli

Print

Grasl Druck & Neue Medien, Bad Vöslau, Austria
Printed on FSC certified paper.

ISBN: 978-3-7001-6968-0

Abe, Karl-Friedrich

Since 1991, Karl-Friedrich Abe has been head of Rhön BR administration in Thuringia. After the political changes in East Germany he strived towards the designation of Rhön as UNESCO Biosphere Reserve, in cooperation with voluntary environmentalists in Hesse and Bavaria.

Artemov, Igor

Igor Artemov is a senior researcher at the Central Siberian Botanic Garden RAS and at Katunskiy BR. His professional interest covers a number of topics related to the flora of the mountains of Southern Siberia, and the conservation and monitoring of vascular species and plant communities.

Aspizua Cantón, Rut

Rut Aspizua Cantón is Forest Engineer by training. She focuses on environmental management, nature conservation as well as forest regeneration and diversification in Sierra Nevada BR. At the moment she is the technical coordinator of the Sierra Nevada Global Change Monitoring Programme, including the monitoring of climate, atmosphere, water and biological parameters.

Badenkov, Yuri P.

Yuri Badenkov is leader of the Mountain MAB-6 Group at the Institute of Geography at the Russian Academy of Science. He coordinates the UNESCO-MAB-6 (mountains) project in Russia since 1983. His research activities focus on sustainable mountain development in the global change context, and the role of biosphere reserves for the conservation of biological and cultural diversity. He is member of the Mountain Agenda group who developed the Mountain Chapter in the Agenda 21. He has about 50 years field experience in research in Central Asia, Caucasus, Altai-Sayan, Far East and many other mountains of North Eurasia. He is member of the UNESCO-MAB International Advisory Committee for Biosphere Reserves (2005-2009).

Bender-Kaphengst, Svane

Svane Bender-Kaphengst holds a diploma in Landscape Ecology and Nature Conservation (Greifswald University, Germany). She specialised in natural resource management, development cooperation and environmental economics, and conducted scientific and feasibility studies on natural resource and water management and environmental education in protected areas in Cuba, Honduras, Guatemala and Albania. After three years of providing expertise in fishery's management for German governmental institutions as well as the European Commission, she focussed her work on the management of protected areas with special reference to UNESCO BRs at Humboldt University Berlin (GoBI). Since 2005 she works with 'The Nature and Biodiversity Conservation Union' (NABU) as international officer, where she became Head of the Africa Programme in 2009. At NABU she is responsible for projects in Africa dealing with the establishment and management of protected areas, forest conservation, poverty reduction and regional development. She initiated and guided the process of the establishment of Kafa BR in Ethiopia.

Bertzky, Monika

Monika Bertzky studied Biology with a focus on Biodiversity and Tropical Ecology (Rheinische Friedrich-Wilhelms-Universität Bonn, Germany). In 2004, she joined the Governance of Biodiversity Project (GoBI), supported by the Robert Bosch Stiftung, investigating the factors of success and failure in BR management. She completed her PhD thesis, which included case studies in the Mexican mountain BRs Sierra Gorda and Sierra de Manantlán, in early

2009. She is a member of the IUCN World Commission on Protected Areas. Since early 2009, she is working as a Programme Officer in the Climate Change and Biodiversity Programme at the UNEP World Conservation Monitoring Centre in Cambridge, UK.

Bomhard, Bastian

Bastian Bomhard holds an MSc degree in Conservation Biology (University of Cape Town). From 2005 to 2009, he worked in IUCN's Programme on Protected Areas, where he was involved in the evaluation and monitoring of natural World Heritage Sites. Since early 2009 he works in the Protected Areas Programme at the UNEP World Conservation Monitoring Centre in Cambridge. There he coordinates work on protected area indicators and targets, protected area management effectiveness, and natural World Heritage Sites.

Bonet García, Francisco

Francisco Bonet García is Biologist by training. Currently he is working at the University of Granada, Spain. He is experienced in database management, data models related to biodiversity conservation, spatial environmental modelling, multi-criteria evaluation techniques applied to natural resource management, and information systems design.

Borowski, Diana

Diana Borowski holds a B.A. in International Relations from the University of Technology Dresden (Germany) and a M.A. in European Interdisciplinary Studies from the College of Europe. Within the scope of her university studies, she focused on EU internal market law and especially on EU measures in the field of environmental and agricultural policies. Her master's thesis dealt with the European Emissions Trading Scheme, exploring the effectiveness of this instrument in combating climate change. Diana joined the Centre for Mountain Studies in July 2010, as part of the GEOSPECS team. This project organises research into areas with "geographic specificities" – such as mountain areas, islands, border areas and others – in order to establish a coherent framework to characterise geographic specificities.

Boussaid, Mohamed

Mohamed Boussaid holds a masters degree in agronomy (National School of Agriculture, Morocco). In 2003, he joined the German Technical Cooperation (GTZ) as a technical advisor on the PRONALCD project (Protection de la Nature et Lutte Contre la Désertification). Last seven years his focus is on natural resources management, forest policy issues, protected areas and the implementation of UNCCD in Morocco. Since 2009 he is involved in the organisation of international conferences in Morocco on biodiversity, climate change and desertification issues.

Dovhanych, Yaroslav

Yaroslav Dovhanych is head of the Zoological Laboratory of the Carpathian BR (CBR). He graduated from the Uzhgorod State University in 1978 with a specialisation in Zoology. Since 1981 he has been working at the Carpathian State Reserve (that was the name of the CBR area before 1992), starting with the position of a forest officer. That time he started his researches of mammals and investigations of ecology of rodents, ungulates and carnivores. He is author of over 80 publications on different zoological and nature protection issues, the participant of numerous international projects where he has acted as either an expert or coordinator.

Flousek, Jiri

Jiri Flousek graduated at the Charles University in Prague in the field of Vertebrate Ecology. He is specialised in birds, small mammals and bats. Since 1981, he is working for the Krkonose National Park Administration in different positions. Between 1994 and 2009 he was deputy director for Nature Conservation and Informatics. Currently he works as ecologist in the park with an additional responsibility for management plans, Natura 2000 Sites, Ramsar sites, the implementation of the Bonn Convention and all issues related to the bilateral Biosphere Reserve.

Grabherr, Georg

Georg Grabherr got his degrees at the University of Innsbruck (1975) where he hold the position of an Assistant Professor at the Institute of Botany with a research focus on the biology of alpine grasslands, and conservation. He was appointed as Professor of Vegetation Science to the University of Vienna in 1986. His research has been concentrated on the classification of vegetation (Austria, and worldwide), climate change effects on mountain plants, and conservation issues. Together with Pauli and Gottfried he founded the Global Observation Research Initiative in Alpine Environments (GLORIA). He is corresponding member of the Austrian Academy of Sciences, where he chairs the Austrian MAB Committee since 2004. He supported the establishment of the Vienna Woods BR and the performance of the Großes Walsertal BR.

Gubko, Victoria

Victoria Gubko is head of the Department of Recreation, Public Relations and International Cooperation. Being a teacher by education she has entered the world of nature conservation in 2004 and actively works in this field till today not only within Carpathian BR, but in NGOs, and ecological projects implemented beyond this institution. Victoria is the field manager for the WWF-DCP project "Conservation and sustainable use of nature resources in the Ukrainian Carpathians" implemented in the Carpathian National Nature Park and the Gorgany Nature Reserve since 2008; member of the WWF-DCP, Ukrainian branch; and head of the local NGO "RakhivEcoTour".

Hamor, Fedir

Fedir Hamor is a recognised ecologist with over 20 years of experience in nature conservation and protected area management. Since 1987 he is director of the Carpathian BR (CBR). He successfully struggled for the extension of CBR's boundaries by almost five times more, gained the inclusion of the reserve to the List of UNESCO BRs (1992), worked out propositions to the Law of Ukraine 'On the nature protection fund of Ukraine', initiated and prepared justification for a number of other ecological laws. He is an honoured officer of Nature Conservation in Ukraine and an honoured citizen of Rakhiv; he is awarded with a national medal 'For the Contributions III', an award of the international contest 'Golden Fortune', the Transcarpathian award for regional development, the honoured Ukrainian tourism activist, and with a Romanian town Viseul-de-Sus.

Hawrysh, Andrew

Andrew Hawrysh graduated from the University of Saskatchewan (Canada) with a Bachelor of Arts in Political Science. He runs a small mixed farm with spouse and children in the Redberry Lake BR – growing organic grains for the edible market and forage for rearing livestock. In addition he works for the Water and Waste Water Branch of the City of Saskatoon for the last 27 years in the Operations and Maintenance Department. He is an active member of the board of directors of the Redberry Lake BR for the last nine years. Currently he holds several positions; vice chair of the board; chair of the agriculture committee; active member in the development of the North Saskatchewan River Watershed Protection Plan; chair of the Redberry Lake Group Environmental Farm Plan; and active member of the Ukrainian Cultural Community within the region of Redberry as well as of Saskatoon.

Kalmikov, Igor

Igor Kalmikov is Wildlife Biologist and graduated at the Wildlife Management Institute in Irkutsk. He worked as wildlife biologist in Chukotka (Far East) and as ranger/manager in protected areas in Eastern Kazakhstan, Sayan mountains (Sayano-Shushenskiy Biosphere reserve). He is coordinator of the Snow leopard monitoring and conservation project (UNDP/GEF Altai-Sayan eco-regional project). Since 2007, he is director of Altaiskiy BR.

Kašpar, Jakub

Jakub Kašpar graduated at the Charles University in the field of Cultural Anthropology. Currently he is deputy director of the Krkonoše National Park Administration. He is responsible for public and international relations since January 2010. Between 2002 and 2009 he worked as director of the Communication Department of the Ministry of the Environment of the Czech Republic. From 1995 to 2002 he worked as journalist specialized mainly in environmental topics and issues.

Köck, Günter

Günter Köck studied Biology at the University of Innsbruck, Austria. His extensive research focuses on biomonitoring studies of aquatic ecosystems. Since 1997 he has been leading projects of the Austro-Canadian research cooperation High-Arctic. In the year 2000 he was awarded the Canada Prize of the University of Innsbruck. In 2004 he became coordinator of the national and international research programmes of the Austrian Academy of Sciences. He is the Austrian delegate to the International Coordinating Council of UNESCO´s Man and the Biosphere Programme and to the European Alliance of Global Change Research Committees, and member of the Scientific Council at the Venice-based UNESCO Regional Bureau for Science and Culture in Europe (BRESCE). Furthermore he is one of the Austrian delegates to the International Scientific Committee for Alpine Research. In 2004, and again in 2010, he was elected as Vice-Chair of the UNESCO MAB Programme. Since 2009 he is co-editor of the scientific journal "eco.mont". In 2010, the Canadian Government recognized his exceptional contributions to Canada-Austria relations by awarding him the Canadian "Go for Gold" honorary medal.

Lange, Sigrun

Sigrun Lange holds a diploma degree in Biology with focus on tropical high mountain ecosystems (University of Bayreuth, Germany), and an MSc degree in Protected Areas Management (University of Klagenfurt, Austria). Since almost 20 years she works in the field of biodiversity conservation and public relation, with field experiences in Papua New Guinea, Kenya, and Ecuador. She was co-organiser of the international congress 'Conservation of Biodiversity in the Andes and the Amazon Basin' hold 2001 in Cusco, Peru. Since seven years she particularly deals with the broad field of protected areas management with focus on BR and transboundary cooperation in protected areas management. In 2005, she coordinated the process of establishing national criteria for BR in Austria. As of 2008, she is CEO of E.C.O. Germany (Munich) specialised on communication, management and planning processes in protected areas. She is honorary communication coordinator of the Alumni Club of the international MSc Programme 'Management of Protected Areas' and also gives lectures in this course.

Maikhuri, R.K.

R.K. Maikhuri is an ecologist with G. B. Pant Institute of Himalayan Environment and Development (an autonomous Institute of Ministry of Environment and Forests, Government of India) focusing on sustainable rural development in the Indian Himalaya.

Messerli, Bruno

Bruno Messerli is a geographer with a special focus on mountain research and development. In 1968 he has been nominated a full professor, in 1978 he became director of the Geographical Institute and between 1986 and 1987 he was rector of the University of Bern, Switzerland. From 1958 to 1976 he conducted fieldwork in the mountains around the Mediterranean Sea and Africa on recent and past glaciation; from 1979 to 1996 he worked on natural hazards and water resources in the Nepal-Himalaya and in Bangladesh, and from 1988 to 1996 on climate change issues in the high Andes of the Atacama region. He held many positions, e.g. director of UNESCO's MAB Programme in the Swiss Alps (1977–1986) or president of the International Geographical Union (1996–2000). He was a founding member of the International Centre for Integrated Mountain Development in Kathmandu 1983, of the African Mountain Association in Ethiopia 1986, and of the Andean Mountain Association in Chile 1991. He was engaged in the preparation of the mountain chapter in Agenda 21 of the Rio Conference 1992 and in the International Year of Mountains 2002.

Munteanu, Catalina

Catalina Munteanu holds a Mag.rer.nat degree in Geography from the Leopold-Franzens-University (Austria) and a degree in Geography and Foreign Languages from the Babes-Bolyai University of Cluj-Napoca (Romania) where she graduated first in her class. Catalina's research focuses on mountain regions (esp. Carpathians and Balkans). Her research and work experience relate to environmental protection, cultural landscapes and development of mountain areas, including GIS teaching assistantships and volunteering. She currently works at the Centre for Mountain Studies (Perth, Scotland) on an international project supporting sustainable development within Europe's mountain regions (mountain.TRIP) and managing a knowledge exchange project in collaboration with the Cairngorms National Park, Scotland.

Nune, Sisay Hailemariam

Sisay Nune is NABU's Ethiopian national coordinator for the project "Climate protection and preservation of Primary Forests in Ethiopia as an Example", which is implemented in the Kafa Zone, Southern Nations Nationalities and People's Regional State, in Ethiopia. He is experienced in natural resources and participatory forest management, establishment of biosphere reserves and closely follows the process of climate change management in Ethiopia. He got his first degree in Forestry from Wondo Genet College of Forestry (Ethiopia) and an MSc degree on Natural Resources Management (specialisation: Forestry for Sustainable Development) from the International Institute for Geo-information Science and Earth Observation (the Netherlands).

Papp, Cristian-Remus

Cristian-Remus Papp holds an MSc degree in Management of Protected Areas from the University of Klagenfurt (Austria). He has over seven years of working experience in the field of nature conservation, out of which five in protected areas related activities. Currently, he coordinates the 'Bears and Ecological Corridors Programme' at WWF Danube Carpathian Programme and is the main collaborator for running the activities of ProPark – Foundation for Protected Areas. He is also a trainer in protected area related fields, project proposals and ecotourism products and services evaluator. He is co-founder and member of several NGOs and member of the Scientific Council of the Maramures Nature Park (Romania).

Pedraza Ruiz, Roberto

Roberto Pedraza is currently working as technical assistant of the Sierra Gorda Ecological Group and as director of the 'Lands for Conservation Programme' which is engaged in buying land for conservation purposes or in establishing payment regimes for ecosystem services. Since several years he has been involved in species conservation activities in Sierra Gorda BR (Mexico), such as inventories of bird or jaguar populations. He is member of 'The Climate Project', a worldwide network of volunteers personally trained by Al Gore to educate and raise awareness about climate change. Besides, he is documenting the biological diversity of Sierra Gorda BR through photography showcased in 'Sierra Gorda, Privilegio de la Patria', and soon in Mexico´s City Chapultepec Zoo in a collective expo about Mexican wildlife with other photographers organised by National Geographic. He just won the 1st place of the 2010 International Year of Biodiversity Photo and Video competition with a photo of a wild margay, which was showcased at the COP10 meeting at Nagoya, Japan.

Pollock, Rebecca

Rebecca Pollock holds a masters degree in Geography from University College London (UK) and recently completed her PhD at Trent University (Canada) in Canadian Studies on the role of UNESCO biosphere reserves in governance for sustainability. She is adjunct faculty at the University of Waterloo in the department of Environment and Resource Studies and teaches a field course about biosphere reserves as complex social-ecological systems. Becky helped to establish the Georgian Bay Littoral BR (Ontario, Canada) in 2004 and is now the communications manager for the Georgian Bay BR Inc. where she coordinates conservation and education projects in support of sustainable development. She is also the past president of the Canadian Biosphere Reserves Association and a volunteer with the Canada MAB committee.

Pokynchereda, Vasyl

Vasyl Pokynchereda is deputy director of the Carpathian BR (CBR). In 1983 he graduated from the Lviv Forest Technical University, and afterwards from the Ecological Faculty of the Timiryazev Academy of Agriculture in Moscow. From 1996 until 2000 he did a postgraduate course at the Institute of Zoology at the Ukrainian National Academy of Sciences. Since 1985 he has been working at CBR in the fields of biological diversity conservation in the Carpathians, in particular with respect to primeval forests. He is author of over 50 publications and one of the authors of the dossier of the Ukrainian-Slovak UNESCO World Natural Heritage nomination 'Primeval Beech Forests of the Carpathians' (2006–2007). He is member of the Zeleni Karpaty editorial board and a founding member of the Ukrainian Centre of Bats Protection. He participated in numerous international projects in the role of a leader, national coordinator or expert.

Price, Martin

Martin Price is director of the Centre for Mountain Studies at Perth College UHI, Scotland. He was appointed professor of Mountain Studies in 2005 and has held the UNESCO Chair in Sustainable Mountain Development since 2009. He played key roles in the formulation and implementation of Chapter 13 of the Agenda 21 on 'Protecting Fragile Ecosystems: Sustainable Mountain Development', endorsed by the Rio Earth Summit in 1992, and the International Year of Mountains, 2002, and has acted as a consultant on mountain issues to international organisations including the European Commission, the EEA, FAO, IUCN, UNDP, UNESCO, and UNEP.

Rao, K.S.

K.S. Rao is a professor of Ecology and Botany at the Delhi University (India), with research, development and teaching interests in ecology, natural resource management and sustainable development.

Rodríguez-Rodríguez, David

David Rodríguez holds a degree in Biology, a postgraduate degree in Conservation Biology (University Complutense of Madrid) and an MSc degree in Ecological Restoration (University of Alcalá de Henares). He has worked in biodiversity conservation for the Spanish Ministry of the Environment (2006) and the European Commission (2008). Since 2008, he develops his PhD thesis on the integrated assessment of protected areas at the Institute of Economics, Geography and Demography of the Spanish National Research Council.

Sarmiento, Fausto

Fausto Sarmiento is an associate professor of Mountain Science in the Geography Department at the University of Georgia, USA. He was President of the Andean Mountains Association (AMA) and Chair in 2002 of the Mountain Geography Specialty Group of the Association of American Geographers (AAG). He works in mountain ethnoecology, forest transitions and farmscape transformation amidst global environmental change. He is the node for the Americas Cordillera Transect (ACT) network of the Mountain Research Initiative (MRI). His professional training includes a bachelor's degree from Catholic University of Ecuador (Biological Sciences), a masters' degree from Ohio State University (Tropical Ecology) and a doctorate degree from the University of Georgia (Landscape Ecology).

Saxena, K.G.

K.G. Saxena is a professor of Ecology and Environmental Sciences in Jawaharlal Nehru University, New Delhi (India), with research, development and teaching interests in ecological perspectives of sustainable development in marginal mountain regions of Asia.

Schmidt, Matthias

Matthias Schmidt is assistant professor for Geography at the Freie Universität Berlin (Germany). He is specialised in human geography, development research and political ecology of mountain areas of South and Central Asia. After studying Geography, Geology, Limnology and Political Sciences at the University of Bonn he did his Ph.D. on Water and Property Rights in Baltistan (Northern Pakistan) with focus on local institutions of resource management. Political and socioeconomic transition processes and their influence on livelihood strategies and natural resource management in Kyrgyzstan are his main fields of interest for the last years.

Shigreva, Svetlana

Svetlana Shigreva holds a degree in Biology from the Altaiskiy State University, Barnaul. She participated in training courses in USA (2005), Austria and Germany (2007, ecological education and PR). Since 2006, she is deputy director of the Altaiskiy Biosphere Reserve and head of the department "Ecological Education/ Training and Public Relations".

Thiel, Lydia

Lydia Thiel is teacher and project leader of coffee K.U.L.T., a coffee brand grown in Sierra Nevada de Santa Marta BR in Colombia and merchandised in Colombia as well as in German-speaking BRs. Her job-related engagement in environmental education and biodiversity conservation induced her to develop the vision of establishing partnerships between biosphere reserves in developing countries and Germany. After searching for appropriate partners she managed to establish contacts with coffee farmers in the Sierra Nevada de Santa Marta BR.

Vološčuk, Ivan

Ivan Vološčuk has been working as director of the Tatry National Park and BR, Slovakia (1990–1996), general director of the Slovak National Park Service (1997–1998), dean of the Faculty of Ecology and Environmental Sciences of the Technical University in Zvolen (1999–2003), and coordinator of the Tatry BR (2004–2007). Since 2008 he is professor at the University of Matej Bel in Banská Bystrica (Institute of Science and Research and Faculty of Nature Sciences). Besides, he holds many positions in non governmental institutions, such as president of the Association of the National Parks and Protected Areas of Slovakia, chair of the Slovak National Committee for IUCN, vice-chair of the Slovak Ecological Society of the Slovak Academy of Sciences, vice-chair of the Slovak National Committee for UNESCO's MAB Programme, member of the IUCN World Commission on Protected Areas and of the Species Survival Commission and Commission on Ecosystem Management. He promoted the acceptance of the Carpathian Primeval Beech Forests in UNESCO´s World Heritage Sites List in 2007. In 1993, he won the WWF Gold Medal.

Yashina, Tatjana

Tatjana Yashina is Landscape Ecologist. She is the deputy director of Katunskiy BR, and has 10 years' experience in coordinating monitoring and research efforts in this protected area. Currently she is also engaged into UNDP-ICI Project 'Extension of protected areas network for conservation of the Altai-Sayan Ecoregion' as coordinator of its climate component.

© ECNC

Zisenis, Marcus

Marcus Zisenis is Senior Officer – Biodiversity Analysis at the European Environment Agency's European Topic Centre on Biological Diversity (ETC/BD) seconded by the European Centre for Nature Conservation (ECNC). He has been working for more than 15 years in nature conservation, landscape planning, and sustainable development in public administration, politics, and research on local, European, and international levels. His research focus is on interdisciplinary evaluations of biodiversity, including methods and societal backgrounds of natural sciences and humanities. Recently, he conducted a biodiversity analysis of major ecosystem types in Europe within the ETC/BD, including mountain ecosystems, which has been published as '10 messages for 2010' by the European Environment Agency.